BIRDS
WINGS & TAILS

BIRDS
WINGS & TAILS

Jack B. Kochan

STACKPOLE
BOOKS

This book is dedicated to the far too few individuals around the world whose life work is the care, pr̶̶̶̶̶̶̶̶̶̶̶̶itation of wildlife.

Published by
STACKPOLE BOOKS
5067 Ritter Road
Mechanicsburg, PA 17055

Printed in the United States of America

10 9 8 7 6 5 4 3 2 1

First edition

Cover design by Caroline Miller and Tina Marie Hill

Library of Congress Cataloging-in-Publication Data

Kochan, Jack B.
    Wings & tails / Jack B. Kochan.
        p.   cm.—(Birds)
    Includes bibliographical references.
    ISBN 0-8117-2503-0
    1. Wood-carving.   2. Birds—Anatomy.   3. Birds in art.   4. Wings.
    5. Tail.   I. Title.   II. Series: Kochan, Jack B. Birds.
TT199.7.K634   1996
731'.832—dc20

                                      95-25138
                                          CIP

Contents

Foreword

When we artists set out to create a realistic piece of art—whether it be a woodcarving, drawing, painting, sculpture, or taxidermy mount—we quickly realize that it is a much more complicated process than we first believed. To succeed, we must build upon a foundation of good research. At the beginning of each project, we must understand the mechanics and technical bases of the parts of the whole. We must have an accurate framework before we add our visual interpretations and emotional responses.

Jack Kochan has given us fine reference material in his series on the anatomy of birds. He has explained heads and eyes, feet and legs, bills and mouths, and now, wings and tails.

In this book, he strips the wings and tail down to their skeletal frames and from there describes the structure of the bones, muscles, and feathers. He goes into great depth on the feather itself, a detail that artists need to understand completely in order to carve or paint any bird realistically. He shows how the wing moves and how it functions as part of the whole. He discusses how wings and tails work together in the complicated process of flight—the key to the bird's survival.

Wings & Tails will enhance your understanding of the anatomy of birds. For artists, research using references like this book, as well as photographs, taxidermy mounts, or live birds themselves, is the most important step in executing a successful project.

Pamyla L. Krausman

Preface

Over the millennia, birds have been artistically depicted in many ways. Bird art has appeared as paintings on the walls of caves, stone carvings, wood sculptures, parchment drawings, and even as part of a written language in Egyptian hieroglyphics. Modern-day wildfowl artists tend to strive for accuracy and realism, and only very recently has impressionistic and interpretive art become popular in the wildfowl carving world.

Both realism and impressionism, though, require a thorough knowledge of the subject. Among carvers and artists, the first bit of bird anatomy that is learned is, perhaps, about the wing. Nearly every carving-related book that is published contains a page or two about the anatomy or topography of the wing. Many of these books give the impression that once you know about primaries and secondaries there is no need to go farther. But these two main feather groups are just a "slice of the pie" compared to the total information that should be gleaned by any serious wildfowl carver or artist.

It was with this thought in mind that this book was written. An artist cannot accurately depict a bird without knowing about the structure, mechanics, and appearance of the species. *Wings & Tails* provides the carver, artist, and bird enthusiast with the details of anatomical structure, mechanics, functions, and appearance of the wings and tails of birds in general.

Although this book is intended as a reference for the wildfowl carver and artist, it is also a valuable source of reference for students and bird enthusiasts.

Wings

ONE

Topography

Topography is the term used by ornithologists to describe the mapping of the surface areas of a bird. This "map" is quite useful when discussing or describing a particular part of a bird. Using common terminology when talking about any part of a bird makes it simple for anyone to understand what is being said.

The topography of the dorsal or top side of the wing.

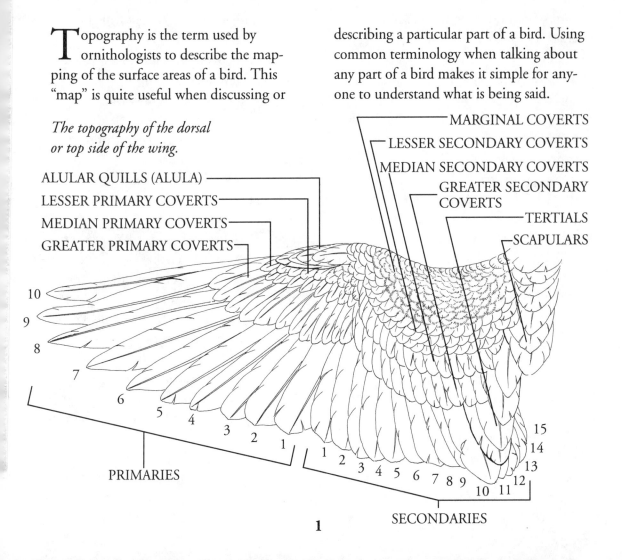

ALULAR QUILLS (ALULA)
LESSER PRIMARY COVERTS
MEDIAN PRIMARY COVERTS
GREATER PRIMARY COVERTS

MARGINAL COVERTS
LESSER SECONDARY COVERTS
MEDIAN SECONDARY COVERTS
GREATER SECONDARY COVERTS
TERTIALS
SCAPULARS

10
9
8
7
6
5
4
3 2 1
1 2 3 4 5 6 7 8 9 10 11 12 13 14 15

PRIMARIES

SECONDARIES

Without common terminology, ornithologists would have to write "the rows of small feathers covering the bases of the primaries" every time they referred to the primary coverts. You can readily see the inconvenience and misunderstanding that this method would produce. Knowing the topography of a wing is critical to understanding the text that follows.

A bird's wing is basically a three-part unit (as described in chapter 2) and contains several feather groups. Looking at the dorsal (top) surface of a wing, you will readily see that the largest feathers are divided into two groups—the primaries and the secondaries.

The primaries are located on the distal (farthest from the body) segment of the wing. In ornithology, they are numbered, starting with the innermost feather. Many times a published work, pertaining to wildfowl art, refers incorrectly to the outermost primary as being number one. This could be somewhat confusing to the novice who is searching for information about a particular species. The tenth, or outermost, primary feather is often much shorter and narrower than the other nine, and in a few species it is absent altogether.

The secondaries are located on the middle segment of the wing and are numbered and counted from the outside in. The primaries and secondaries, collectively, are called remiges and vary in number between different species. The most common number of primaries is ten, with extremes of nine and eleven. The secondaries vary from eight to as many as forty, as on the albatross. Eleven or twelve is the most common number.

Covering the bases of the primaries, on the dorsal side of the wing, is a row of feathers called the greater primary coverts. There is one covert for each primary feather. A second row of shorter coverts overlaps the row of greater primary coverts; these feathers are called the median primary coverts. Not all species have this row of median coverts. On most species there is a third, even shorter, row of coverts overlapping the second row, and these are called the lesser primary coverts. The common pigeon does not have this third row of lesser primary coverts.

The secondaries are overlapped at their bases by a row of feathers called the greater secondary coverts, and unlike the primaries, there is not always one covert for each secondary feather.

In some groups of birds, the fifth secondary feather is absent, leaving a slight gap between the fourth and sixth secondaries. There is, however, a fifth secondary covert present. This gap in the secondaries is called a diastema, and birds having this gap are said to be diastataxic. When there is no gap, and the fifth secondary feather is present, the birds are known to be eutaxic.

A second row of shorter coverts over-

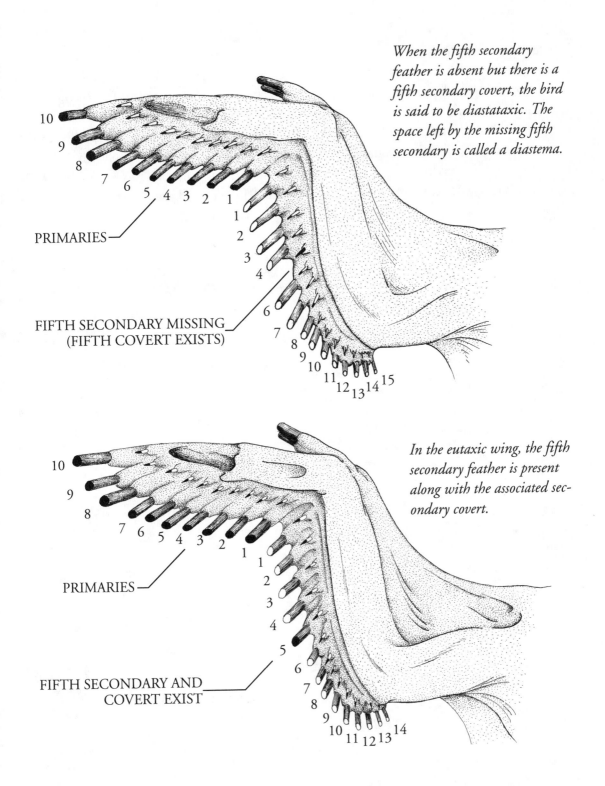

When the fifth secondary feather is absent but there is a fifth secondary covert, the bird is said to be diastataxic. The space left by the missing fifth secondary is called a diastema.

PRIMARIES —

FIFTH SECONDARY MISSING
(FIFTH COVERT EXISTS)

In the eutaxic wing, the fifth secondary feather is present along with the associated secondary covert.

PRIMARIES —

FIFTH SECONDARY AND
COVERT EXIST

laps the row of greater secondary coverts; these feathers are called the median secondary coverts. Overlapping the median secondary coverts are one to three rows of even shorter feathers called the lesser secondary coverts. The lesser secondary coverts are not always distinguishable as being definite rows of feathers. All the remaining coverts on the dorsal surface of the wing are known as the marginal coverts.

Located at the anterior (front) edge of the distal (outermost) segment of the wing is a group of prominent feathers called the alular quills. There are normally three alular quills, but there can be from two to five depending on the species. Collectively, the alular quills are called the alula, and

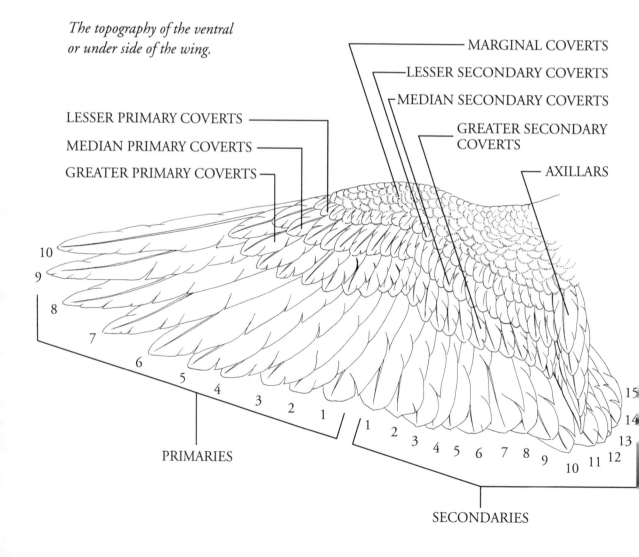

The topography of the ventral or under side of the wing.

MARGINAL COVERTS

LESSER SECONDARY COVERTS

MEDIAN SECONDARY COVERTS

GREATER SECONDARY COVERTS

AXILLARS

LESSER PRIMARY COVERTS

MEDIAN PRIMARY COVERTS

GREATER PRIMARY COVERTS

PRIMARIES

SECONDARIES

they are very important in the mechanics of flight. The alula is sometimes colloquially referred to as the bastard wing.

On the proximal (closest to the body) segment of the wing is a group of feathers called the tertials. The tertials are sometimes also referred to as the tertiaries or humeral feathers.

The tertial feathers are actually a continuation of the secondaries but are not considered to be remiges. They are relatively long, and often quite different in shape from the secondaries. The tertials serve the purpose of closing the gap between the body and the wing during flight. Located on the shoulder of the bird is a group of feathers called the scapulars.

The ventral (under) surface of the wing has the same feather groups as the dorsal surface. The bases of the primaries are covered by a row of greater primary coverts and, once again, there is one covert for each primary feather. The second and third rows of primary coverts are likewise called the median primary coverts and the lesser primary coverts.

As on the dorsal surface, the secondaries on the underside have the same rows of coverts and are labeled the same— greater secondary coverts, median secondary coverts, and lesser secondary coverts. The remaining area of the ventral surface is again covered by the marginal coverts. The coverts on the ventral side of the wing are usually much more alike and are less easily distinguished than the coverts of the upper wing. These underside coverts are often referred to as the wing lining.

On the proximal segment, inboard to the secondaries, is a feather group that corresponds to the tertials on the upper surface. Growing from the armpit area, these feathers are called the axillaries, or axillars, and are usually longer and stiffer in size and shape than the underwing coverts. The axillars are normally shorter than the tertials but serve the same purpose of closing the gap between the body and the wing during flight. The remaining area of the wing, both top and bottom, is covered by the marginal coverts.

TWO

Skeletal System

The skeletal structure of a bird's wing is more easily understood when compared to the skeleton of a human arm. Since both skeletal systems are similar, they are said to be homologous. Like the arm of a human, the wing is divided into three segments.

The innermost segment contains the humerus, which is attached, at the proximal end, to the pectoral girdle, or shoulder. The distal (outermost) end of the humerus forms a joint with two other bones—the ulna and the radius. This joint is the elbow.

The middle segment of the wing containing the ulna and radius forms a joint, at their distal ends, with the manus. This joint is the wrist, or bend of the wing, and the manus corresponds to the hand of a human. The wrist joint contains two other small, somewhat squarish, bones— the radiale and the ulnare. These two bones correspond to the carpals of the human wrist.

The segment of the wing containing the humerus, between the body and the elbow, is called the brachium, while the middle segment, between the elbow and wrist, is called the antebrachium or forearm. The outermost segment, beyond the wrist, is the manus or hand. The manus is distinctively different from the human hand in that some metacarpals are missing while others have fused together to form the carpometacarpus bone.

Only three fingers, or digits, are found on the manus of a bird. The bones that make up the digits, or fingers, of the manus are called phalanges. Identifying the digits of the manus is slightly different from identifying the toes. On a bird's foot, the toes are numbered (digit I, digit II, etc.), but on the manus they are given names. The first digit, containing only

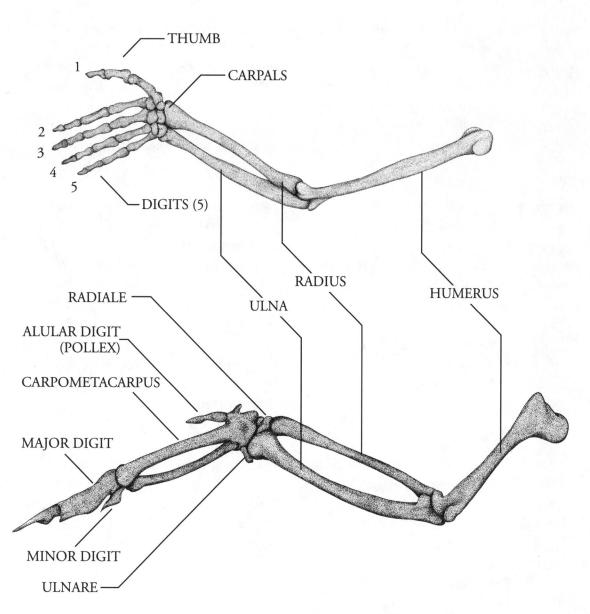

THUMB

CARPALS

1

2
3
4
5

DIGITS (5)

RADIALE

ALULAR DIGIT
(POLLEX)

CARPOMETACARPUS

MAJOR DIGIT

MINOR DIGIT

ULNARE

ULNA

RADIUS

HUMERUS

*The skeleton of a bird's wing is comparable
to the arm of a human. In birds, the
metacarpal bones have fused to form the
carpometacarpus, and a bird has only three
digits instead of five as in humans.*

one phalanx (singular of phalange), is the alular digit, which compares to the human thumb. The alular digit has some freedom of movement and articulates to move the alula. In early ornithology, the alular digit was called the pollex, a term still sometimes found in use today.

The other two digits of the manus are the minor digit, having one phalanx, and the major digit, which typically has two phalanges. It is generally believed, but not proven, that the major digit corresponds to the second digit, or index finger, of the human hand, while the minor digit corresponds to the third digit, or middle finger, of the human hand.

The proximal phalanx of the major digit is very flat with a sharp edge at the rear, while the single phalanx of the minor digit is somewhat triangular in shape. In a few species, such as the hoatzin, there is a claw on the end of the alular digit and sometimes on the major digit. These wing claws are used for tree climbing and are probably relics of the birds' reptilian ancestors.

The major digit, minor digit, and carpometacarpus bone are the support for the primary flight feathers. The ulna is the thicker of the two bones of the forearm and is the support for the secondary flight feathers. Usually there exists on the posterior edge of the ulna a series of bumps. Known as quill knobs, these bumps are the points where the secondaries are attached.

Movement of the wing segments is limited to specific planes of motion, mostly because of the shape of the bones at the joints. A joint is defined as the point where two or more bones meet, and there are two kinds of joints. The articulated joint is one where the two adjoining bones move with respect to each other, creating a bend in the skeletal structure. Most of the articulated joints are also synovial. A synovial joint is totally encapsulated within a thin tissue layer and is filled with a fluid. This synovial fluid is the lubricant for the joint.

A second type of joint is the suture, where two or more bones meet but do not move with respect to each other. Except for the alular digit and the major digit, the bones of the manus form joints that are basically sutures. The joints of the manus do not move up and down. These joints can bend only inward, toward the body, thus keeping the manus in a flat plane.

The section of the skeletal system that is seldom thought of as part of the wing is the pectoral girdle. The pectoral girdle is comprised of three separate bones that form a tripod to support the humerus. Knowledge of the pectoral girdle and its association with the wing is important in understanding how the wing moves. The enlarged proximal end of the humerus is the head, which fits into a socket on the pectoral girdle. This socket is called the glenoid cavity, and is formed by the meet-

The pectoral girdle forms the socket for the head of the humerus and creates the shoulder of the bird.

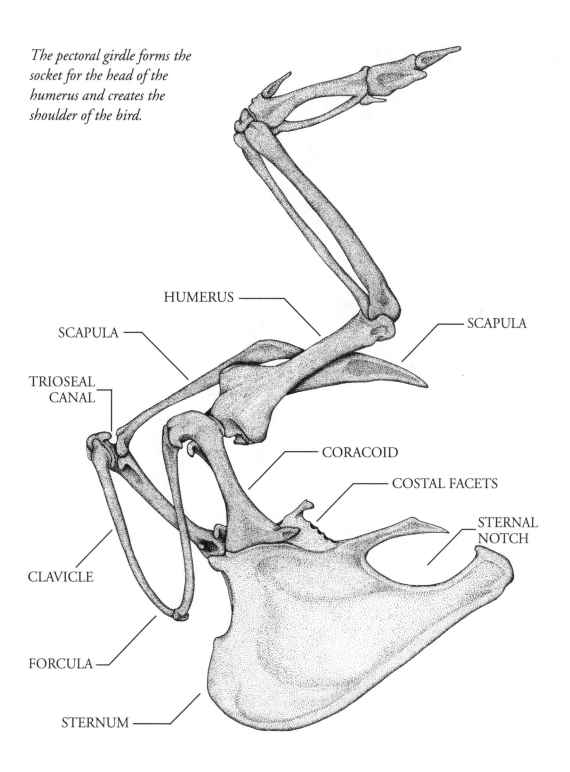

HUMERUS

SCAPULA

SCAPULA

TRIOSEAL
CANAL

CORACOID

COSTAL FACETS

STERNAL
NOTCH

CLAVICLE

FORCULA

STERNUM

ing of three bones—the coracoid, the scapula, and the clavicle—of the pectoral girdle. The junction of the coracoid and scapula forms the socket for the head of the humerus, while the joining of the clavicle forms an opening called the foramen triosseum. (See chapter 3.)

Although the glenoid cavity is a ball and socket joint, the humerus is somewhat restricted in movement because the head of the humerus is not spherical in shape. The humerus has basically up-and-down motion and also can bend backward toward the body to fold the wing. Unlike the human upper arm, rotation of the humerus is very restricted and basically moves in only two planes—up and down, front and back.

The coracoid extends from the front end of the scapula to the sternum (breastbone), and forms a strong brace for the wing. The scapula is a rather long, thin bone that extends rearward, where it attaches to the ribs. In penguins, the scapula is broad and rather heavy and is more like the human scapula.

The two curved clavicle bones (one on each side of the bird) extend forward and downward from the shoulders toward the sternum (breastbone). The lower ends of the clavicles are fused together to form the forcula (wishbone). The forcula acts as a spring to hold the pair of wings apart. Although the forcula is considered to be important for flight, it is modified in some

parrots and owls. In these birds, the lower ends of the clavicles are not fused together but are joined instead by flexible cartilage. The clavicles are relatively much smaller in flightless birds and are totally absent in the kiwi.

As with the humerus, there is little or no rotary motion of the forearm (radius and ulna). The forearm has only front-and-back motion and limited up-and-down movement. This front-and-back movement allows for the folding of the wing and, along with the manus, forms a firm, flat support for the flight feathers. The wrist joint flexes forward and backward, up and down, but cannot rotate like the wrist of a human.

The skeletal system of a penguin's wing is somewhat different than in flying birds. The humerus is much shorter and thicker, with two small sesamoid bones at the elbow. Likewise, the radius and ulna in penguins are shorter, broader, and much more flattened. The ulna does not show signs of having quill knobs. The carpals (radiale and ulnae) of the penguin are replaced with the scapholunare and the pisolunare, the latter of which is very broad, flat, and triangular shaped. There is no alular digit, and the metacarpals are also broad and flat.

The bones of birds and animals are held together at the joints by strong, cord-like, fibers called ligaments. Ligaments are attached to adjoining bones and act like

The skeleton of a penguin wing differs somewhat from the wing of other birds.

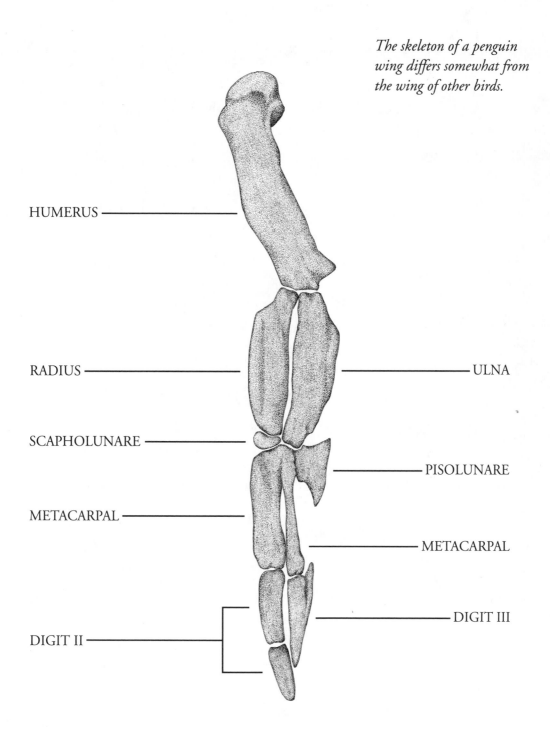

HUMERUS

RADIUS

ULNA

SCAPHOLUNARE

PISOLUNARE

METACARPAL

METACARPAL

DIGIT III

DIGIT II

ropes or chains to bind the bones together so that they do not stretch when pulled and collapse when pushed.

Knowing the mechanics of the wing skeletal system is important when carving or depicting a bird's wing. The carver or artist must know the limitations of skeletal movement in order to accurately reproduce the wing.

THREE

Muscles

The skeleton is the framework struc-ture of any animal or bird, and the muscles are the motor devices that provide movement of this structure. Without muscles the skeleton would be totally inanimate and would lack the form that is produced by muscle mass.

The muscles of a bird are of three basic types: cardiac, striated, and smooth. Of the wing, we are concerned only with the striated muscles, since the cardiac muscles are associated with the heart and circulatory system and the smooth mus-cles are found in the skin. There are more than two dozen separate muscles associ-ated with the wing, and these are more easily understood if we look at them in two groups—the ventral, or underside, group and the dorsal, or topside, group.

Learning the scientific names of the muscles is a somewhat discouraging task for someone not particularly interested in myology, which is the study of muscles

and the muscular systems. Learning what the muscles do is a bit easier if you under-stand how or why a muscle was named.

Most muscles have a Latin name refer-ring to their task or size. For example, a flexor muscle causes the joint to bend, while an extensor muscle causes the joint to straighten. A pronator muscle causes a segment to rotate about the joint, while an abductor muscle causes an appendage to be moved away from the longitudinal centerline of the body. Muscles in the same group are often named with refer-ence to their size—major and minor (large and small) or longus and brevis (long and short). Thus, the flexor digitorium and the extensor digitorium are two muscles that bend or straighten a digit.

Understanding the muscle system of the wing is somewhat easier if we consider the muscles in two groups—the ventral muscles and the dorsal muscles. Nearly all the muscles of the wing are located close to the body and give the bird a more aerodynamic shape.

Most of the muscles on the ventral side of the wing are flexor muscles serving to elevate and depress the wing or flex the forearm. The largest, most powerful ventral muscle of the wing is not even located on the wing. In fact, many people do not even think of this as a wing muscle. The breast contains the pectoralis, which is the largest and most powerful muscle of any bird capable of flight. Located beneath the pectoralis, next to the breastbone, are the supracoracoideus and the coracobrachialis. These three muscles of the breast are the primary actuators that elevate and depress

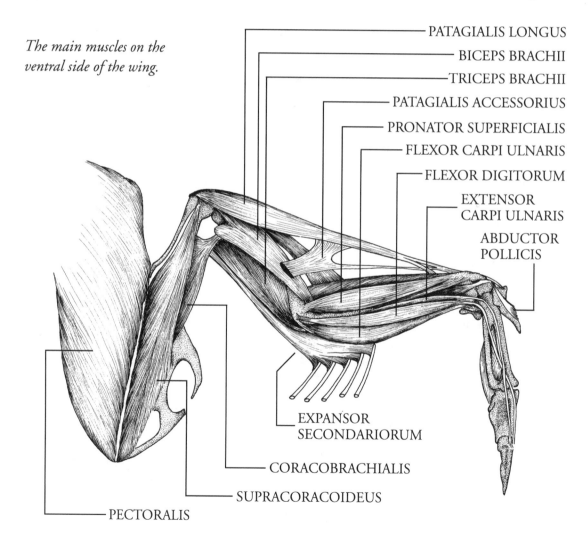

The main muscles on the ventral side of the wing.

PATAGIALIS LONGUS

BICEPS BRACHII

TRICEPS BRACHII

PATAGIALIS ACCESSORIUS

PRONATOR SUPERFICIALIS

FLEXOR CARPI ULNARIS

FLEXOR DIGITORUM

EXTENSOR CARPI ULNARIS

ABDUCTOR POLLICIS

EXPANSOR SECONDARIORUM

CORACOBRACHIALIS

SUPRACORACOIDEUS

PECTORALIS

(flap) the wing for flight. The muscles of the breast make up from 30 to 60 percent of the total weight of a bird.

Other ventral wing muscles serve to spread the secondaries (expansor secundarium), move the digits (flexor digitorium), or extend the hand (extensor metacarpi).

Most of the remaining muscles of the wing are located on the two innermost segments. There are no muscles located on the manus or hand segment except for a small muscle that extends the alula. Movement of the digits is accomplished by muscles found on the forearm, or middle segment, through a unique cable-and-pulley system.

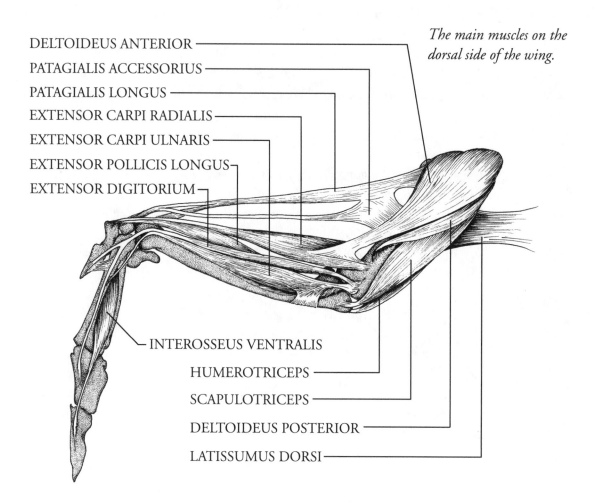

DELTOIDEUS ANTERIOR
PATAGIALIS ACCESSORIUS
PATAGIALIS LONGUS
EXTENSOR CARPI RADIALIS
EXTENSOR CARPI ULNARIS
EXTENSOR POLLICIS LONGUS
EXTENSOR DIGITORIUM

INTEROSSEUS VENTRALIS
HUMEROTRICEPS
SCAPULOTRICEPS
DELTOIDEUS POSTERIOR
LATISSUMUS DORSI

The main muscles on the dorsal side of the wing.

Muscles are attached to the bones by tendons, which are strong sinuous fibers found only at the ends of a muscle. A muscle can do work only by pulling, not pushing, much like a stretched rubber band. The breast muscle supracoracoideus, for example, terminates with a tendon that passes over the coracoid bone and attaches to the upper surface of the humerus. When the supracoracoideus contracts, or pulls, the cable-and-pulley system causes the humerus to raise, thus elevating the wing. The pectoralis is the muscle that powers the lowering, or downstroke, of the wing.

The muscles of an articulated joint are always found in pairs. For each muscle that causes a movement there is a companion muscle that causes the opposite movement, such as flexor and extensor muscles. When the flexor muscle is working, or pulling, the companion extensor muscle is relaxing. When both these muscles are pulling at the same time they hold the joint in a fixed position, as in the spread wing.

Muscles are made up of stringy fibers arranged in bundles. These muscle bundles are held together by a thin tissue called facia. There are also layered sheets of facia between muscle groups that separate them and bind them together. All muscles contain both red fibers and white fibers. Because the white fibers tire more easily, in birds that fly well the breast muscles contain mostly red fibers. In birds that do not fly for great distances, such as the domestic chicken, the breast muscles have more white fibers. These red and white fibers are also what give us the light or dark meat in any poultry.

Each muscle of a bird is represented on both sides of the bird. For example, there are two pectoralis muscles (one for each wing), one on each side of the sternum (breastbone). These two muscle masses lying side by side leave a depression between them. When carving the breast area of a bird, this depression can be emphasized to portray the tremendous power available, especially of a bird in flight.

Knowing where the muscles are located and what they do can help you produce a more accurate and lifelike sculpture or portrait. Since most of the muscle mass on a wing is located close to the body, when the wing is folded at rest there is a prominent bulge of muscle mass in the scapular or shoulder area. Even on the extended wing there is considerable fullness of the humerus and forearm segments.

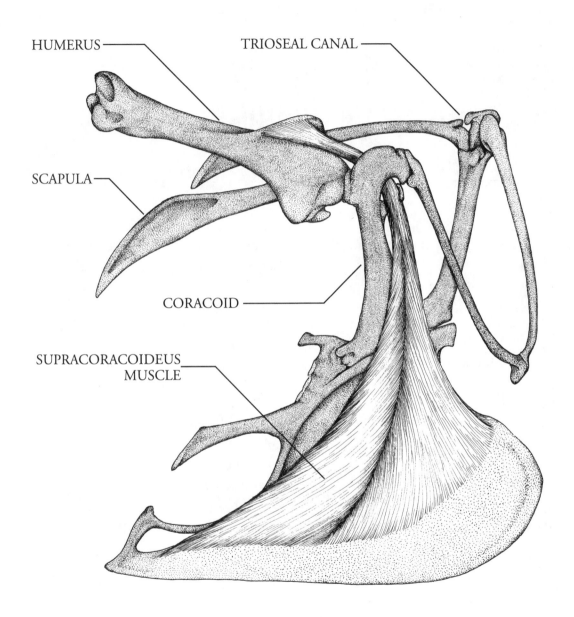

HUMERUS

TRIOSEAL CANAL

SCAPULA

CORACOID

SUPRACORACOIDEUS
MUSCLE

*The breast muscles are the prime
movers of flapping motion. This
movement is accomplished by a
cable-and-pulley system.*

FOUR

Covering

Covering and protecting the bones and muscles of all animals and birds is the integument. The integument includes the skin, scales, claws, bill, and fur or feathers. The skin is basically a two-layer structure made up of the epidermis (outer layer) and the dermis (under layer). The skin is attached to the muscles by a layer of facia tissue.

The dermis contains blood vessels, nerves, nutrient cells, smooth muscle, and fatty tissue. Feathers receive nutrients and a blood supply needed for growth from the dermis.

The epidermis is much thinner than the dermis and contains a protein material called keratin, which is necessary for the production and growth of feathers, scales, and claws. In humans, keratin is what makes up the hair and fingernails. The socket from which a feather grows is called the feather follicle.

The integument of a bird does not have pores or sweat glands as found in humans because birds do not perspire.

Most species, however, do have an oil gland in the skin that produces an oily substance used for preening. The primary purpose of the skin is, of course, to cover and protect the underlying muscle and tissue.

Located also in the skin, feathers, scales, claws, and bill of a bird are color pigments called carotenoids. These pigments, named after the common carrot, are what give color to the skin and feathers. The blood vessels in the dermis also aid in giving color to the skin and sometimes cause the skin to change color similar to a human blushing. This color change is quite visible on certain species, especially during mating season, and primarily in the male.

The smooth muscles found in the skin allow a bird to move its feathers, usually as a group, but some feathers can be moved individually. A bird can fluff, or raise, its feathers or depress them, flattening them against the body. Feathers can also be rotated to some extent around the axis of the feather shaft. This rotation ability is quite pronounced on the primaries and is one of the many functions that takes place during flight.

Along the anterior, or leading edge, of the wing is a rather loose fold of skin that extends from the distal end of the humerus to the wrist. This fold of skin is called the patagium. Inside of this skin fold are the tendons of the biceps slip muscle. When the wing is spread, these tendons are stretched tight, giving a taut support for the skin. This action creates a thin, sturdy, leading edge on an already aerodynamic wing. A second, smaller fold of skin also exists from the distal end of the humerus to the trunk, or body. This smaller fold of skin is the humeral patagium and is what creates the axilla (armpit).

HUMERAL PATAGIUM ———

BICEPS SLIP ———

PATAGIUM ———

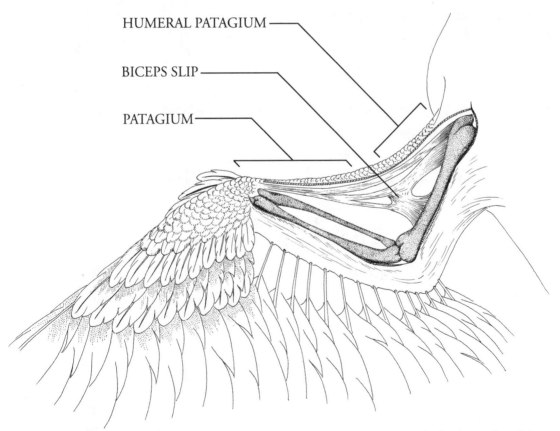

The patagium is created by a tendon running through a fold of skin on the leading edge of the wing. The biceps slip muscle is the divider for the patagium and humeral patagium.

The feather tracts and regions on the dorsal side of the wing and tail.

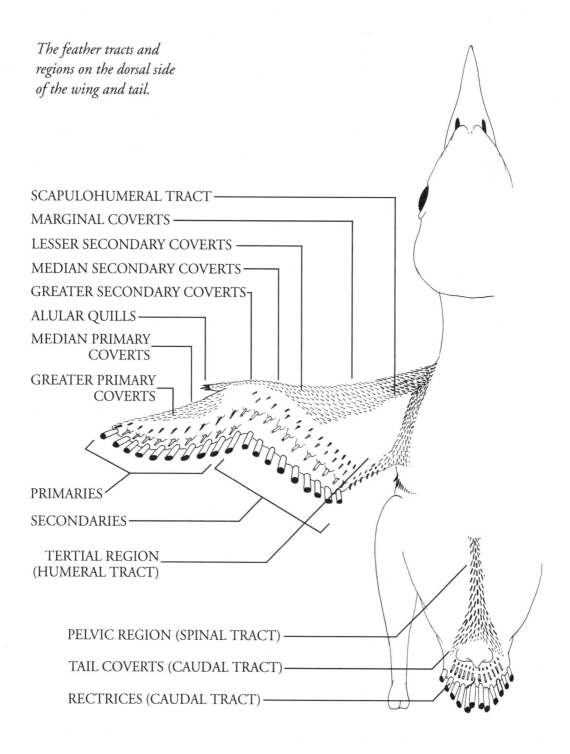

SCAPULOHUMERAL TRACT

MARGINAL COVERTS

LESSER SECONDARY COVERTS

MEDIAN SECONDARY COVERTS

GREATER SECONDARY COVERTS

ALULAR QUILLS

MEDIAN PRIMARY COVERTS

GREATER PRIMARY COVERTS

PRIMARIES

SECONDARIES

TERTIAL REGION (HUMERAL TRACT)

PELVIC REGION (SPINAL TRACT)

TAIL COVERTS (CAUDAL TRACT)

RECTRICES (CAUDAL TRACT)

FIVE

Feather Tracts

As can be seen in the illustration of topography, feathers do not grow just randomly on the wing of a bird. Nor are the feathers distributed evenly over the wing or body, except in a few species, such as penguins and toucans. Feathers grow in definite patterns on a bird, and this is especially true on the wing. These patterns, known as feather tracts, appear all over the body. Although feather tracts vary somewhat between families of birds, they remain fairly constant among all birds.

Feather tracts of the wing are described and named by varying methods by different ornithologists. Some authorities consider the wing as having several separate tracts, while other sources name the wing as a single tract with several regions. Whatever method is used, the results are the same, and the wing is subdivided into more areas, groups, and regions than any other part of the bird.

For simplicity we can consider the wing as having only two feather tracts on the dorsal side and two feather tracts on the ventral side, with each of these tracts having several regions.

In ornithology, a feathered area is referred to as pteryla, while the featherless areas of the skin between the tracts are called apteria. The apteria are not entirely void of feathers and almost always show signs of down feathers. In aquatic birds, such as ducks, the apteria are often heavily feathered with down as a means of insulating the body against the cold water. The apteria generally are not seen on the surface, since they are covered by the contour feathers of the adjoining pterylae (plural).

On the top side of the wing are the scapulohumeral tract and the alar tract. The scapulohumeral tract is a narrow band of feathers found on the shoulder of a bird. This tract begins on the anterior (front) edge of the scapula, or shoulder, and extends obliquely along the brachium (humerus) to the elbow. The scapulohumeral tract contains the scapular feathers. Some sources refer to this tract as being two separate tracts—the scapular tract and the humeral tract—but since there are no apteria between them, they appear as one tract.

On the posterior edge of the humeral segment is the tertial region, considered by some authorities as being a separate tract. This region begins at the proximal end of the humerus near the front edge, where it barely touches the scapulohumeral tract, and extends to the elbow. The tertial region contains the tertials. The tertials are considered by some as being remiges, but they differ from the primaries and secondaries in several respects.

All the remaining feathers on the top of the wing are located in the alar tract. This includes the alular quills, primaries, secondaries, and coverts. As described in the topography of a wing, the feathers in the alar tract are divided into specific groups.

The underside of the wing also contains the humeral tract and the alar tract. The humeral tract on the underside contains the axillary region, located in the armpit area. Once again, this region is considered by some to be the axillar tract. The axillaries arise from this region.

Feathers within a feather tract grow in specific patterns, generally in overlapping rows, and also have a specific angle of orientation. The flight feathers (primaries and secondaries) grow only at the edge of the feather tract, while the coverts grow over the entire tract. Only in a few species, such as the penguin, are the feathers evenly distributed over the entire body.

All the feathers visible on a bird are considered as contour feathers, including the primaries and secondaries. Flight feathers (primaries and secondaries), growing only at the edges of a feather tract, collectively are called remiges.

The primaries all arise in the segment of the manus and point at an angle toward the distal (outer) end. On many species the outermost primary is greatly reduced in size. This small-sized primary is called the remicle, or little remix.

The primaries that are attached to the fused metacarpal bones are referred to as metacarpal primaries. The remaining primaries are referred to as digital primaries because they are attached to the digits. The number of metacarpal primaries serves as a fundamental characteristic in placing a species in the proper order or family. Grebes, storks, and flamingos have seven metacarpal primaries, while nearly all other birds have six.

A term sometimes used for the hand segment, or the primaries collectively, is pinion. Occasionally, the primaries of one wing on waterfowl are purposely removed,

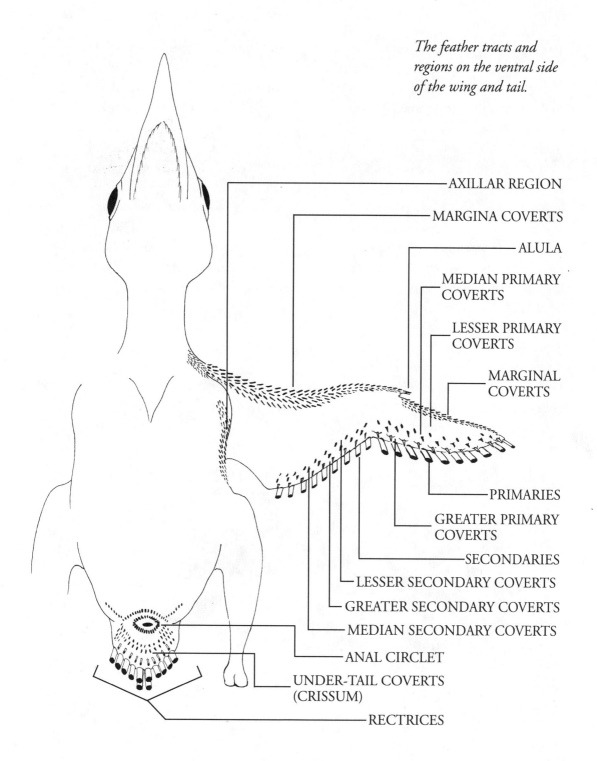

The feather tracts and regions on the ventral side of the wing and tail.

AXILLAR REGION

MARGINA COVERTS

ALULA

MEDIAN PRIMARY COVERTS

LESSER PRIMARY COVERTS

MARGINAL COVERTS

PRIMARIES

GREATER PRIMARY COVERTS

SECONDARIES

LESSER SECONDARY COVERTS

GREATER SECONDARY COVERTS

MEDIAN SECONDARY COVERTS

ANAL CIRCLET

UNDER-TAIL COVERTS (CRISSUM)

RECTRICES

along with the manus, to assure that the birds will stay at a particular pond or site. This practice is called pinioning and fortunately is seldom done anymore. Birds that have been pinioned will never be able to fly.

The secondaries all arise from the middle segment—the antebrachium, or forearm. The secondaries are attached to the ulna and grow at an angle slightly inward. The secondaries are technically called cubitals, from a Latin word meaning elbow. In a few species, there is an additional, smaller, secondary feather located in the space between the primaries and secondaries. This carpal remix seems to be disappearing through evolution but is still present in some gallinaceous birds and gulls.

The alular quills are a group of feathers attached to the pollex, or thumb. The alular quills are numbered from the innermost to the outermost. The alula vary in count from as few as two in hummingbirds to as many as five or six in certain cuckoos.

The primaries all grow from the manus, or hand segment. They are counted and numbered from the innermost to the outermost. Most birds have ten primaries. The alular quills arise from the alular digit.

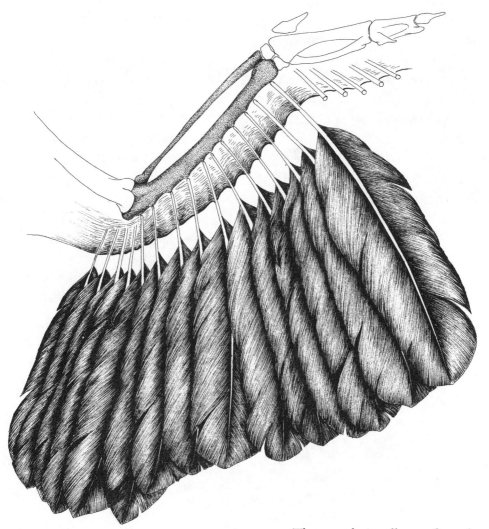

The secondaries all grow from the area of the ulna. They are counted and numbered from the outermost to the innermost. The secondaries vary in number with different species from as few as eight to as many as forty.

The tertials and axillars grow at an angle nearly parallel to the trunk, or body, and the scapulars grow at a slightly outward angle. The upper wing coverts emanate from the skin nearly perpendicularly but have an extreme curvature so that they appear to lie flat. This extreme curvature of the coverts can be emphasized in a carving, either individually or in feather groups.

Feather tracts vary slightly between species, but the feather tracts of the wing are the same on all species capable of flight. Generally, the feathers within a tract remain in their relative position regardless of how the wing is moved. As the wing is spread open, the feathers will remain approximately at the same angle with respect to the segment to which they are attached.

On the folded wing of a bird, the feathers are not just bunched together randomly. As the wing folds, the flight feathers are neatly stacked, atop one another,

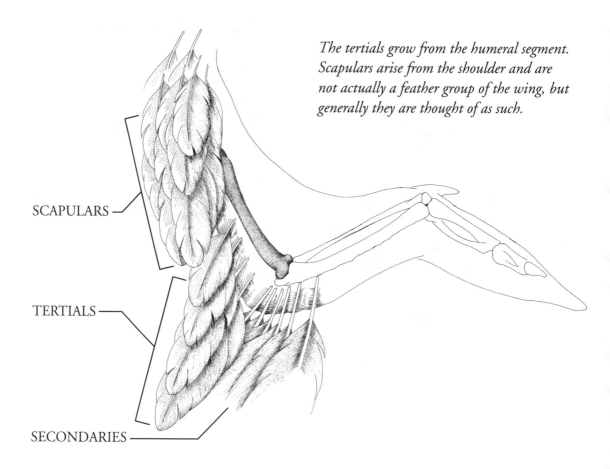

The tertials grow from the humeral segment. Scapulars arise from the shoulder and are not actually a feather group of the wing, but generally they are thought of as such.

SCAPULARS —

TERTIALS —

SECONDARIES —

1

2

3

As the wing folds, the
primaries are neatly stacked
(1), then overlapped by the
also stacked secondaries (2).
The secondaries are partially
covered by the tertials and
scapulars in the folded wing
(3). The groups of feathers
each have a specific placement
on the folded wing.

and form a compact mass next to the trunk or body. The primaries are neatly stacked and tucked under the secondaries, which are also stacked with the innermost secondaries tucked under the scapulars and tertials. This stacking of feathers and the associated muscle mass creates a prominent bulge at the shoulder and along the scapular area. This mass should be a prime consideration when carving or illustrating any wildfowl.

SIX

Size and Shape

While studying a species, an ornithologist will take many measurements of the bird. These measurements are important in determining rate of growth, differences between sexes, and variation within the species. Two basic wing measurements that are always taken are labeled (W.) and (Ex.). Measurements of a bird are usually recorded in millimeters, though sometimes they are shown in inches.

The wing (W.) measurement is taken in a straight line from the wrist joint, or bend of the wing, to the tip of the longest primary feather when the wing is folded. Note that the longest primary is not necessarily the outermost feather. The extent (Ex.) is a measurement of the spread wings, from tip to tip of the longest primary feathers of the two wings.

On a live bird, measuring the extent is

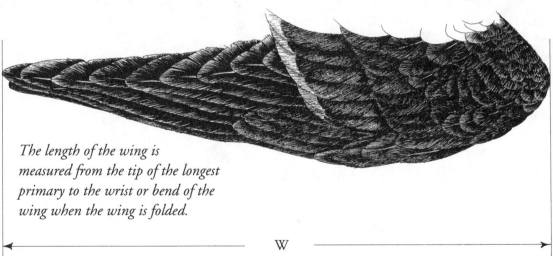

The length of the wing is measured from the tip of the longest primary to the wrist or bend of the wing when the wing is folded.

W

accomplished by laying the bird on its back and grasping the two wrist joints with the thumbs and forefingers. The wings are gently spread open and the measurement taken. On large birds, such as the Canada goose, two people may be required to safely and properly spread the wings. This handling does not harm the bird and provides the ornithologist with valuable information without having to destroy the specimen, as was done in the early days of ornithology.

The extent, or wing-spread, is measured with the wings held open while the bird is on its back. Measurement is from tip to tip of the longest primaries.

Other measurements that are sometimes taken are the lengths of each primary, secondary, and alular quill.
Sometimes the distance from the tip of a
primary to the tip of its corresponding
greater covert on the folded wing is also
measured. Whenever these additional
measurements are available, they are a
valuable aid to the carver or artist.

The wings of birds do not, of course,
all have the same shape and size. The
shape and size of the wing determine, to a
great extent, what style of flying a bird
does or if the bird can fly at all. Many
hawks and gulls, for instance, will use
soaring flight whereas passerines normally
flap their wings to stay aloft.

When the wing measurement (W.) is
decidedly longer than the trunk or body
of the bird, as in terns, the wing is classed

*In the rounded wing, the middle
and innermost primaries are the
longest and decrease in length
toward the outermost primary,
as in the wing of a grouse.*

as long, whereas a short wing is one where W is the same as, or less than, that of the trunk length, not including the head and tail, as in the grebe.

In a rounded wing, the middle primaries are the longest, while the remaining primaries are graduated, or each one successively shorter than the one next to it, as seen on the wing of a quail. On the pointed wing, the outermost primaries are the longest, as in a Canada goose.

The pointed wing of a Canada goose has the outermost primaries being the longest with the innermost primaries being the shortest.

A wing is said to be broad when the primaries and secondaries are relatively long, as in the broad-winged hawk. When the primaries and particularly the secondaries are relatively short, the wing is classed as narrow, as in a gull.

The relatively long primaries and secondaries form a broad wing on the broad-winged hawk.

With relatively short primaries and secondaries, the wing of a tern is considered to be narrow.

All wings have a longitudinal curvature, though some are curved more than others. If the curvature is extreme, as in the grouse, it is said to be concave, whereas the wing of a hummingbird or swift is only slightly curved and is said to be flat.

The concave wing of a grouse has an obvious curvature.

A common mistake of the novice carver portraying a spread wing is to make the wing rather flat. This gives the appearance of the bird having a board stuck to its body. A spread wing has quite a bit of shape in more than one plane, and every carver should study many references to gain knowledge about wing shape.

One other characteristic of a wing that is rarely seen is that of a peculiar bony structure on the bend of the wing. This structure is in the shape of a spur, and the wing is said to be spurred, as in the jacana and the spur-winged plover of Africa.

The individual feathers in a group will often vary in shape, and this is especially true of the flight feathers. The most notice-able variations occur on the primaries. The primaries are asymmetrical, in that the shafts of the primaries are never located down the exact longitudinal centerline of the feather. The outermost primary feather has the quill, or shaft, located very close to the forward, or leading, edge. The shafts then are located progressively closer to the centerline on each primary. The tenth, or outermost, primary is usually much nar-rower than the next primary inboard.

The primaries also vary in shape on most birds, and this variation is most con-spicuous at the tip of the feather. The out-ermost primaries are often modified in shape in that they are narrowed, for about one-third of their length, at the tips.

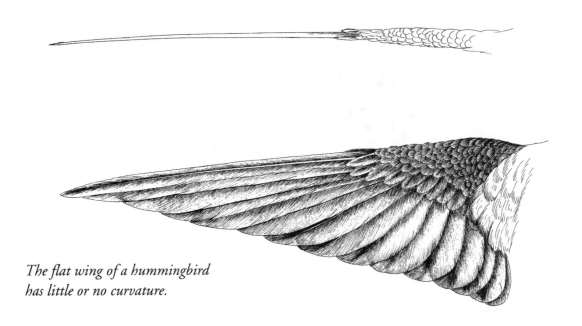

The flat wing of a hummingbird has little or no curvature.

These narrow tips create what are called slots in the open wing. Slots are a valuable asset to certain types of flight, especially high-speed maneuverability.

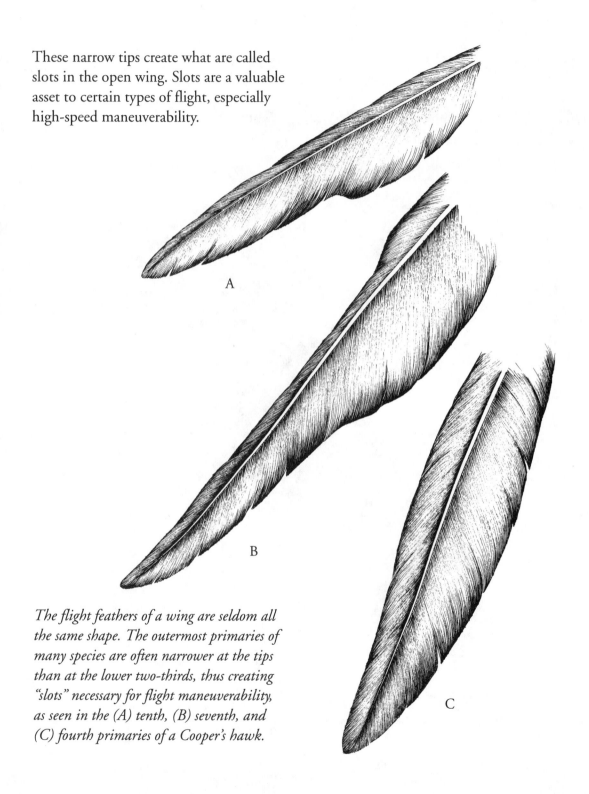

The flight feathers of a wing are seldom all the same shape. The outermost primaries of many species are often narrower at the tips than at the lower two-thirds, thus creating "slots" necessary for flight maneuverability, as seen in the (A) tenth, (B) seventh, and (C) fourth primaries of a Cooper's hawk.

The variations of feathers on a wing have to do primarily with flight. In most owls, for example, the outermost primaries have a rather soft, comblike leading edge. Also, the trailing edges of an owl's pri- maries are softer than those normally found in primary feathers. These soft, more flexible feather edges reduce the noise made during flight, giving the predator a silent advantage over the prey.

The outermost primaries of owls have a comblike leading edge that aids in silent flight.

The secondaries of most birds are also asymetrical. The shafts become closer to the longitudinal centerline on each feather, progressively inboard toward the body. On birds with rather broad secon-daries, the inner vanes of the feathers usu-ally have ripples. These rippled vanes, when carved into a wing, can create a very realistic representation.

Broad flight feathers usually have a rippled inner vane as seen here on a secondary feather of a Canada goose.

Wing shape is one of the primary means for bird-watchers to identify different types of hawks, falcons, and eagles. Accipiters are a group of hawks that share common characteristics and include the goshawk, sharp-shinned hawk, and Cooper's hawk. Their wings are short and rounded, and their tails are usually long and slender.

Buteos are another group of hawks that include the redtail, red-shouldered, Harlan's, rough-legged, ferruginous, and others. This group has wings that are long, broad, and rounded. Buteos normally soar in continuous circles to search for prey.

Harriers are not a group of hawks like buteos and accipiters are. Harrier is a name given only to the marsh hawk because of its hunting tactics. The harrier will usually fly close to the ground in search of food.

Falcons fall into a group called falco, and all except the caracara have long, narrow, pointed wings that are usually held in a sweptback position during a high-speed dive while attacking prey. The caracara has long, somewhat rounded wings.

Eagles are generally larger than hawks, and also have long, broad, rounded wings. Eagles normally soar at a higher altitude than hawks. Owls are readily distinguished from hawks and falcons by their unique body and wing shape, as well as their hunting tactics.

Wing shape is a means of identifying hawks, falcons, and eagles in flight. (A) Long, broad, rounded wings and a rounded tail identify buteos. (B) Short, rounded wings, along with long, narrow, squarish tails are typical of accipiters. (C) The pointed, sweptback wings let you quickly identify a falcon. (D) Owls are readily distinguished by their compact body and their wing shape. (E) Eagles have long, broad, rounded wings, are larger than hawks, and generally soar at a higher altitude.

SEVEN

Position and Use

Flying, of course, is the most obvious use of a bird's wings. Active flapping flight occurs only in four of the animal groups: insects, pterosaurs (extinct), bats, and birds. A bird uses its wings, however, for many other purposes. The second most commonly observed use is, perhaps, in courtship displays done primarily by the male of a species.

Nearly everyone is familiar with the drumming of the ruffed grouse. This sound, like the staccato of a snare drum, is produced by the grouse beating its wings very fast, causing the air to be displaced. This displacement of the air makes a popping noise or a snap that is repeated many times in succession. It was once thought that this noise was produced by the tips of the primary feathers of the opposing wing tips hitting each other, but high-speed photography has disproved this belief. During courtship, the woodcock uses its primary feathers to produce a low, whistling sound while it flies.

Most hawks and owls will spread their wings in a draped fashion over the prey they have just killed. This protective, covering action is called mantling. The mantling display is sometimes also used as a defensive attitude against a predator. The draped wings give the appearance that the bird is much larger than it really is, thus discouraging a would-be attacker.

Mantling is a display used by most birds of prey.

The killdeer, some plovers, and a few other species use their wings as a distraction. As a predator nears the nest of a killdeer, the bird will almost invariably move off the nest with one or both wings drooped and dragging on the ground. This action causes the predator to believe that the bird is injured and will be easy prey. The bird continuously draws the attacker away from the nest, and just when the enemy thinks there is a sure capture the bird will fly off to safety, leaving the predator a good distance from the nest. Unfortunately, this distraction tactic does not work every time.

The killdeer uses an injured wing display to lure predators away from the nest.

Wings are also used as defensive weapons, as you would readily learn if you venture too close to a nesting Canada goose or a swan. These large birds can deliver quite a disturbing blow using only their wings as a defensive tool.

Not all birds use their wings for flying. Some species have lost, or never attained, the ability to fly. The ostrich is a flightless bird, and its lack of flying ability is compensated for by its ability to run fast and well. The penguin, however, can neither fly nor run. In fact, it has great difficulty even in walking. The penguin is an excellent swimmer, though, and uses its wings as a pair of flippers to propel itself through the water. Some species of diving ducks also use their wings to "fly" underwater. In many species the wing becomes a safe haven for the chicks during times of danger, required warmth, or other needed protection.

As previously mentioned, the wings of many species are used in courtship displays, and also in similar displays before and after copulation. These displays, called precopulatory and postcopulatory displays, are markedly different from mating displays. The Canada goose, for example, immediately after copulation will raise its breast out of the water, lay its neck back, point its bill upward, and partially spread and arch its wings. This position is displayed by both the male and female

and is followed by an immediate stretching of the wings.

At least one species uses its wings as an aid to food gathering. The black heron of Africa will spread and arch its wings while it stands in shallow water. The spread wings form an umbrella to cast a shadow on the water where small fish seek cooling shelter and are quickly caught by the bird.

Most birds will sunbathe by spreading their wings (and tails). This action apparently helps to warm the bird and also helps rid it of parasites. Anhingas and cormorants perform a similar action known as wing drying.

All of the above described actions of the different species are called behavior patterns. The study of behavior is called ethology. Behavior patterns occur at different levels. The lowest level of behavior is a motor act, such as flapping flight or soaring, and is called an action pattern.

There are eight main categories of behavior patterns. The copulatory displays would be included in the sexual behavior pattern, while preening is included in the maintenance behavior pattern. The other patterns are spatial, nutritional, agnostic, nesting, parental, and interspecific.

The mechanics of the folding of the wing were discussed in chapter three. The folded wing, however, requires special attention by the carver and artist. The position of the folded wing is not the same on all birds. Waterfowl, for instance, hold the folded wing quite differently than do passerines. Even within the passerines the wings are positioned differently between species.

In waterfowl (ducks, geese, swans) the folded wing fits into a special pocket at the upper limits of the side feathers. The side feathers forming the wing pocket partially cover and hide part of the wing. Since these birds spend much of their life on the water, this pocket provides better shedding of water than if the wing were totally exposed.

Many songbirds will hold the folded wing along the sides of the body. This is usually the wing position of birds having a concave wing, as seen in wrens, while others fold the wing with the primaries resting on the rump. In birds with rather long primaries, the wing tips will often cross each other when the wings are folded, as seen in terns and many ducks. When the primaries are extremely long, as in gulls and terns, and the wing is folded, the tips are positioned well above the rump and tail. The extremely concave wing of a grouse is positioned tightly against the entire area of the body.

When depicting a species, the artist or carver should make a thorough study of wing position. How many primaries are exposed, how many secondaries, how many rows of coverts, and how many alula exist? These are all important considerations when carving or illustrating a bird realistically.

EIGHT

Special Adaptations

In most birds, the wing itself is already a special adaptation used for flight. The shape and size of the wing normally determine in what manner the bird flies—soaring, gliding, or flapping. Fast-flying birds, like falcons and swallows, have long, narrow wings, while soaring birds usually have extremely long, broad wings, as can be seen in buzzards. Several other special features that are worth mentioning exist in the wings of birds.

Research by osteologists (those who study bones) show us that the bones of a bird are unique in that they contain many air spaces. This pneumatized structure makes the bones very lightweight yet strong, a necessary requirement for flight. Pneumaticity of the bones is a feature found only in birds. With the many air cavities in the bones, weight is kept at a minimum. The bones of a bird weigh about 20 to 30 percent less than the bones of mammals. In some species, the bones of the skeleton weigh less than the feathers. These air spaces also serve additional functions.

The respiratory (breathing) system of a bird has, in addition to the lungs, numerous air sacs. These air sacs are used to store, moisten, and sometimes filter the air that is breathed. One of these ventilation storage areas is the humerus bone of the wing.

In the females of most species, calcium is built up and stored in the pneumatic cavities of some bones. This calcium build-up is stored in the hollow air spaces of several different bones and produces a secondary bone within a bone. This secondary bone structure is called medullary bone. The medullary bone is later converted back to calcium and used in producing eggshells for incubation. It is believed that the hormone estrogen is what accounts for this process. Males of a species do not produce medullary bone, since estrogen is a hormone found only in the female.

Typical of nearly all bones of a bird, this cutaway illustration of a humerus shows the many hollow spaces used in the respiratory system. These cavities are also used for calcium storage by the female of many species.

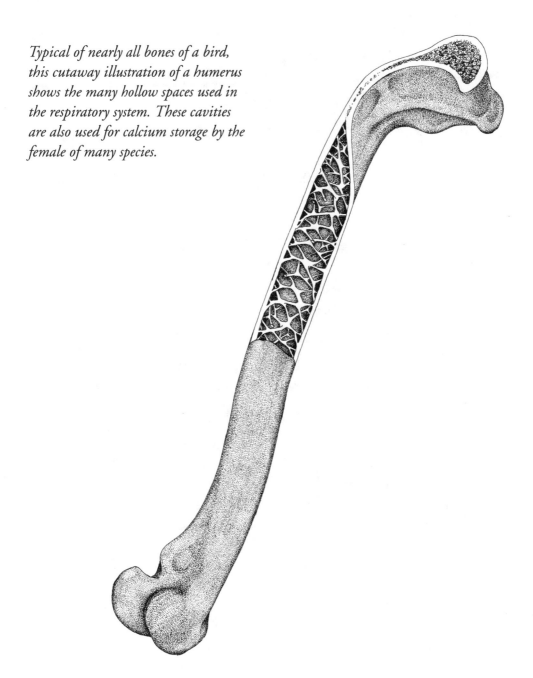

Tails

NINE

Skeletal System

At the mention of a bird's tail, most people visualize a group of flight feathers, lying in a flattened plane, projecting rearward at the extreme posterior (rear) end of the bird. This common image is only part of the entire picture.

The tail of a bird is, in reality, composed of several parts, only one of which is the flight feathers. The tail of a bird is actually a small bone structure covered with muscle, flesh, and feathers. The bone structure of the tail includes the caudal

The skeleton of a bird's tail is made up of six caudal vertebrae and a pygostyle.

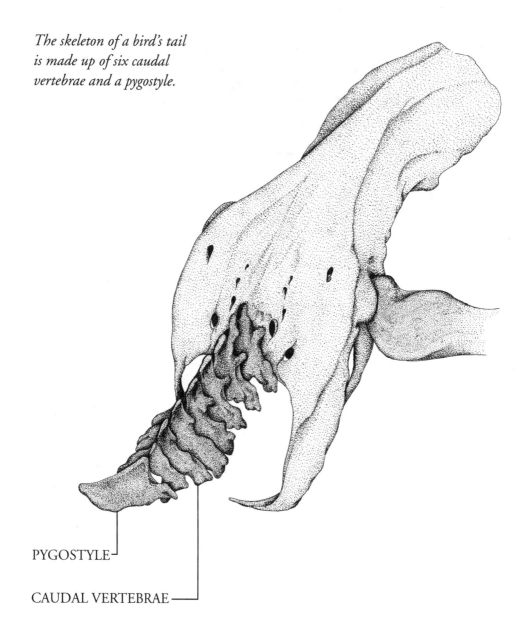

PYGOSTYLE⌐

CAUDAL VERTEBRAE ⌐

vertebrae and the pygostyle. There are six caudal vertebrae in birds, and these allow for the flexibility of the tail. The final few vertebrae are fused into one flat blade of bone that supports the muscles and surrounding tissue. This blade is the pygostyle. Together, the six caudal vertebrae and the pygostyle make up the caudal section of the spinal column. The pygostyle is roughly ploughshare-shaped and is the base for the rectrices, or flight feathers of the tail.

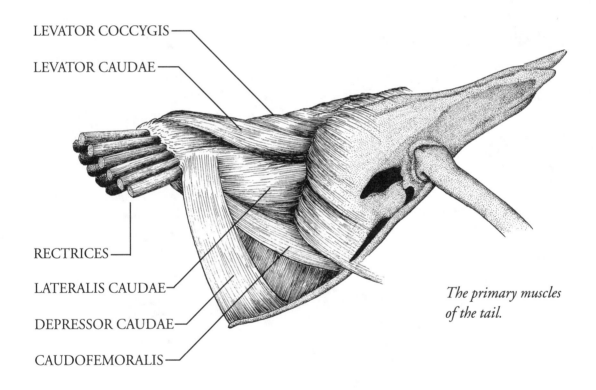

LEVATOR COCCYGIS

LEVATOR CAUDAE

RECTRICES

LATERALIS CAUDAE

DEPRESSOR CAUDAE

CAUDOFEMORALIS

The primary muscles of the tail.

Located also on the dorsal side of the tail, above the pygostyle, is the uropygial gland. The more common terminology for this gland is the oil gland, or preen gland. The uropygial gland produces an oily substance that is used in preening the feathers. Not all species have a uropygial gland. When the uropygial gland exists, it is generally a relatively large structure that terminates in a nipplelike opening. Depending on the species, there can be from one to eight openings on the gland. There is sometimes a cluster of slender feathers surrounding the nipple that create a brush for dispensing the oily secretion.

The uropygial gland is largest in aquatic birds. The oil gland does not exist, however, in ostriches, rheas, emus, and cassowaries. Ducks and other waterfowl have a rather large oil gland whose secretion is used to coat the feathers. It was once thought that this oily coating was what made the feathers of a duck water repellant. Research has shown that this is not necessarily true.

In birds that had their uropygial gland surgically removed at birth, the feathers were just as water repellant as in birds of the same species that still had an oil gland. Without this oily coating though, the feathers lose much of their efficiency in their normal functions of flight and insulation and tend to wear out more quickly. Ornithological experiments appear to indicate that the oiling of the feathers is done primarily to maintain the quality of the feather rather than waterproof it.

Oil is not what makes the feathers

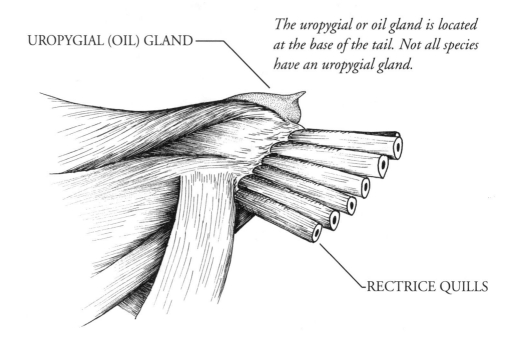

UROPYGIAL (OIL) GLAND

The uropygial or oil gland is located at the base of the tail. Not all species have an uropygial gland.

RECTRICE QUILLS

waterproof. In fact, excessive oil causes the feathers to mat more easily, thus reducing the insulating quality. Waterfowl that have become covered with oil in an oil spill will usually die from thermal exposure.

The topography of the tail contains only three areas: the rectrices, the upper tail coverts, and the under tail coverts.

The rectrices are paired; there are an equal number on each side of the body center-line. Most birds have six pairs of rectrices (twelve), but hummingbirds, swifts, and most cuckoos have five pairs (ten). A few species have an extremely high number of rectrices. The ring-necked pheasant has nine pairs and the white pelican has twelve

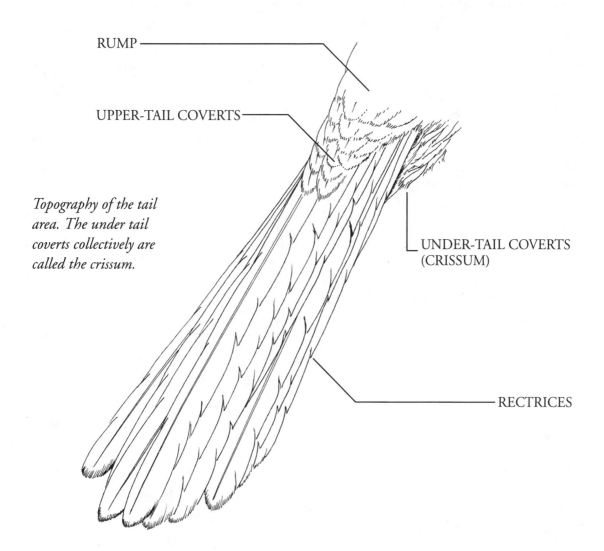

RUMP

UPPER-TAIL COVERTS

UNDER-TAIL COVERTS (CRISSUM)

RECTRICES

Topography of the tail area. The under tail coverts collectively are called the crissum.

pairs. Ostriches are said to have more than fifty rectrices.

The upper tail coverts are often indistinguishable from the feathers of the rump and are not as clearly defined in rows as they are in the coverts of the wing. Since they appear very much like the feathers of the rump, an imaginary line must be drawn across the bird directly above the vent in order to determine the extent of the upper tail coverts and the beginning of the rump area.

The under tail coverts cover the bases of the rectrices on the ventral (lower) side and extend to an imaginary line drawn across the bird through the vent. Collectively, the under tail coverts are known as the crissum. All of the feathers of the tail lie in the caudal tract.

TEN

Size and Shape

The size, or length, of a bird's tail is a measurement that is always taken by ornithologists. This measurement is recorded as T. in the measurement charts and is the length of the longest pair of rectrices from their tip to the point where they attach at the base. Tail size varies with each species and is considered to be either short or long. A short tail is one where the length is the same as or less than the length of the wing measurement W. A long tail measures greater than the length of the wing.

Tails come in a variety of shapes and sizes and are classified as to their shape or the formation of the rectrices. Each shape can have slight variations between species, but generally the tail shapes are divided into six different types, depending upon their external characteristics.

The square tail has rectrices all ending at the same length, thus giving the end of the tail a squarish appearance, as in the nuthatch.

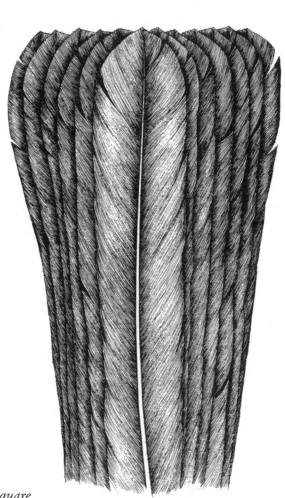

The rectrices of the square tail all end at the same length, as in the tail of a nuthatch.

A rounded tail has rectrices that shorten successively from the center to the outside in slight gradations, giving the tail a rounded shape, as in the common crow.

The rectrices of a crow's rounded tail shorten successively from the center to the outside.

In a graduated tail, the rectrices shorten successively from the center to the outside in abrupt gradations, as in the black-billed magpie.

When the rectrices shorten in abrupt steps from the center to the outside, the tail is gradated, as in the tail of a black-billed magpie.

When the middle rectrices are much longer than the others, as in the mourning dove, the tail is said to be pointed or acute.

The middle rectrices of the mourning dove are longer than the others, creating a tail that is pointed or acute.

Sometimes the middle rectrices are the shortest and increase in length successively toward the outside pair in slight gradations. This tail is said to be emarginate, as in the snow bunting.

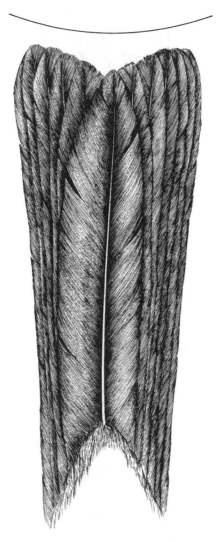

The emarginate tail has the shorter rectrices in the center and gradates successively to the outside, as in the snow bunting.

In the forked tail, the rectrices increase in length from the center to the outside pair in abrupt successive gradations, as in a tern or swallow.

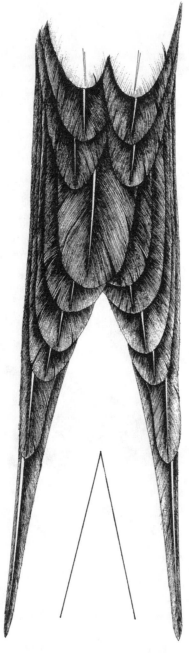

The forked tail of a royal tern has short rectrices in the center and gradates in abrupt steps to the outside.

These six characteristic tail shapes are useful to ornithologists when they catalog different species. Tail sizes and shapes are generally one of the identifiers for classifying a family of birds. For example, the family Sittidae (nuthatches) has a tail (T.) measurement that is much shorter in length than the wing (W.) measurement is square in shape, and the rectrices are broad with rounded tips.

The individual rectrices also have various shapes that can be found in any of the six tail shapes. The chimney swift has rectrices that are spinose. The shaft of the feather is extended without barbs on the outer edge and forms a pointed tip, while the rectrices of a woodpecker end in a very sharp point and are said to be acuminate.

Sometimes the feathers are very wide and are classed as broad.

Feathers on any bird are not a permanent structure. Feathers fray and wear from abrasion and sometimes get bent or broken. Worn feathers of the wings and tail are periodically replaced through a process called molting. All birds go through this molt process at least once a year, and some species molt two or three times a year. No species, of course, sheds all its feathers at one time. Of the flight feathers of both the wing and tail the molting is very systematic. If all the primaries were shed at one time the bird could not fly. This happens only in a few species, and during that time the bird is rather defenseless. Normally, the flight

The spinose feather has the quill, or shaft, extending beyond the barbs at the tip.

feathers are shed one or two at a time, in a specific sequence, so that the bird still has the capability of flight.

In most species, molting coincides with the mating season and is referred to as the breeding plumage or nuptial plumage. The other major molt produces what is called the winter plumage.

Size and shape of both the wing and tail are important characteristics when ornithologists place a species into a family. The following chart shows the characteristics of wings and tails of the various bird families.

The acuminate rectrices of woodpeckers end in a sharp point.

Nuthatches have broad rectrices with squarish tips.

WING AND TAIL CHARACTERISTICS OF BIRD FAMILIES

Albatross (Diomedeidae)
Wing: Long; narrow.
Tail: Short to moderately long.

Anhinga (Anhingidae)
Wing: Moderately long.
Tail: Long; rounded; middle pair of rectrices fluted.

Avocet and Stilt (Recurvirostridae)
Wing: Long; pointed; tertiaries greatly lengthened.
Tail: Short; somewhat square.

Blackbird, Oriole, and Meadowlark Subfamily (Icterinae)
Wing: Varies in length and shape; usually long and pointed.
Tail: Rounded; varie in length; never longer than wing.

Bushtit (Aegithalidae)
Wing: Rounded; ten primaries.
Tail: Longer than wing; rounded and graduated.

Cardinal, Buntings, and Grosbeak Subfamily (Cardinalinae)
Wing: Varies in length and shape.
Tail: Varies in length and shape.

Cormorant (Phalacrocoracidae)
Wing: Short; rounded.
Tail: Long; rounded.

Crane (Gruidae)
Wing: Long; broad.
Tail: Short.

Creeper (Certhiidae)
Wing: Long; rounded; ten primaries.
Tail: As long as wings; rounded; rectrices acuminate.

Crow, Magpie, and Jay (Corvidae)
Wing: Long, pointed in ravens; short, rounded in magpies and jays.
Tail: Rounded; Sometimes graduated.

Cuckoo, Roadrunner, and Anis (Cuculidae)
Wing: Varies in shape and size.
Tail: Long; graduated.

Dipper (Cinclidae)
Wing: Short; rounded; concave; ten primaries.
Tail: Short; square; nearly rounded; rectrices broad and rounded at tip.

Dove and Pigeon (Columbidae)
Wing: Long; flat; varies in shape.
Tail: Long; may be square, pointed, or rounded.

Duck, Goose, and Swan (Anatidae)
Note: Ducks, geese, and swans are divided into several subfamilies and tribes.
Wing: Pointed or rounded.
Tail: Short; usually rounded.

Eagle, Hawk, and Kite Subfamily (Accipitrinae)
Wing: Long; broad; rounded, except kites, which are long, narrow, and pointed.
Tail: Short; varies in shape; usually rounded.

Falcon (Falconidae)
Wing: Long; narrow; pointed.
Tail: Short to moderately long; squarish; nearly forked.

Finch and Grosbeak (Fringillidae)
Wing: Varies in size and shape; nine primaries.
Tail: Varies in size and shape; usually square or slightly forked.

Flamingo (Phoenicopteridae)
Wing: Short; rounded; narrow.
Tail: Short; slightly rounded.

Flycatcher (Tyrannidae)
Wing: Outermost primaries usually longer than secondaries.
Tail: Usually square; sometimes forked.

Frigate bird (Fregatidae)
Wing: Long; pointed.
Tail: Long; deeply forked.

Gannet (Sulidae)
Wing: Long; pointed.
Tail: Long; pointed.

Goatsucker (Caprimulgidae)
Wing: Long; pointed.
Tail: Varies in size and shape.

Grouse and Ptarmigan Subfamily (Tetraoninae)
Wing: Short; concave; rounded; primaries usually curved.
Tail: Varies in size and shape.

Gull, Tern, and Skimmer (Laridae)
Wing: Very long; narrow; pointed.
Tail: Varies in length and shape.

Heron and Bittern (Ardeidae)
Wing: Long; broad; rounded.
Tail: Short.

Horned Lark (Alaudidae)
Wing: Long; pointed; nine primaries.
Tail: Shorter than wing; nearly square.

Hummingbird (Trochilidae)
Wing: Long; flat; pointed; secondaries extremely short.
Tail: Varies in size and shape.

Ibis (Threskiornithidae)
Wing: Long; broad; rounded.
Tail: Short.

Jacana (Jacanidae)
Wing: Long; somewhat rounded; spurred; tertiaries extremely long.
Tail: Short.

Kingfisher (Alcedinidae)
Wing: Long; pointed.
Tail: Long; slightly rounded.

Limpkin (Aramidae)
Wing: Short; rounded; tertiaries and outer primary attenuate and curved.
Tail: Short.

Loon (Gaviidae)
Wing: Short; somewhat pointed.
Tail: Short; stiff rectrices.

Merganser and Eider (Tribe Mergini)
Note: Mergansers and eiders are members of the duck family (Anatidae) and are considered to be diving ducks.
Wing: Varies; usually long; broad; rounded.
Tail: Varies in shape; short to moderately long.

Nuthatch (Sittidae)
Wing: Long; pointed; ten primaries.
Tail: Short; almost square; rectrices broad with rounded tips.

Osprey Subfamily (Pandioninae)
Wing: Long; narrow; pointed.
Tail: Moderately long; rounded.

Owls (Strigidae)
Wing: Varies in size and shape with from one to six of the outer primaries notched.
Tail: Varies in length; usually rounded; rarely square.
Note: Barn owl is in the family Tytonidae. Wings are long, pointed, and inner edge of all primaries are not notched.

Oystercatcher (Haematopodidae)
Wing: Long; pointed.
Tail: Short; square; slightly rounded.

Parrot and Macaw (Psittacidae)
Wing: Varies in shape and size.
Tail: Varies in shape and size.

Pelican (Pelecanidae)
Wing: Very long; rounded.
Tail: Very short.

Pheasant Subfamily (Phasianinae)
Wing: Short; rounded; concave.
Tail: Long; pointed.

Plover (Charadriidae)
Wing: Long; pointed; tertiaries greatly extended.
Tail: Short.

Puffin and Auk (Alcidae)
Wing: Moderately long to short; pointed.
Tail: Short.

Quail Subfamily (Odontophorinae)
Wing: Short; rounded; concave; primaries curved.
Tail: Short; rounded.

Rail, Coot, Gallinule (Rallidae)
Wing: Short; rounded; tertiaries often as long as primaries.
Tail: Short.

Sandpiper, Snipe, and Woodcock (Scolopacidae)
Wing: Long; pointed; tertiaries greatly extended.
Tail: Short.

Shearwater and Fulmar (Procellariidae)
Wing: Long; narrow.
Tail: Short.

Shrike, American (Laniidae)
Wing: Short; rounded; ten primaries.
Tail: Varies in size from short to very long; varies in shape; square, rounded, or graduated.

Skimmer (Rynchopinae)
Wing: Very long; narrow; pointed.
Tail: Forked.

Skua and Jaeger Subfamily (Stercorariinae)
Wing: Very long; narrow; pointed.
Tail: Shorter than wing; rounded or pointed.

Sparrow, Old World (Passeridae)
Wing: Long; pointed; ten primaries.
Tail: Shorter than wing; nearly square.

Spoonbill and Ibis (Threskiornithidae)
Wing: Long; broad; rounded.
Tail: Short.

Starling (Sturnidae)
Wing: Long; pointed; ten primaries; outermost primary acuminate.
Tail: Short (half the length of wing); square to emarginate.

Storm petrel (Hydrobatidae)
Wing: Long; narrow.
Tail: Short to moderately long.

Swallow (Hirundinidae)
Wing: Long; pointed; nine primaries; secondaries very short.
Tail: Not longer than wing; forked or emarginate.

Swift (Apodidae)
Wing: Long; flat; pointed; secondaries extremely short.
Tail: Varies in shape; forked or emarginate; rectrices sometimes spinose.

Tananger Subfamily (Thraupinae)
Wing: Moderately long; nearly pointed.
Tail: Shorter than wing; square to rounded; sometimes emarginate.

Tern (Sterninae)
Wing: Very long; narrow; pointed.
Tail: Forked.

Thrasher and Mockingbird (Mimidae)
Wing: Varies in length, usually short; rounded; ten primaries.
Tail: Varies in length, usually longer than wing; rounded, sometimes graduated.

Thrush Subfamily (Turdinae)
Wing: Long; pointed.
Tail: Usually shorter than wing; square to slightly rounded.

Titmouse (Paridae)
Wing: Rounded; ten primaries.
Tail: As long as wing; slightly rounded.

Towhee Subfamily (Emberizinae)
Wing: Varies in size and shape; usually rounded.
Tail: Extremely variable in size and shape.

Trogon (Trogonidae)
Wing: Short; rounded; concave.
Tail: Long; usually graduated; rectrices broad.

Tropic bird (Phaethontidae)
Wing: Long; pointed.
Tail: Pointed; middle rectrices are filamentous (long, slender, without barbs most of its length).

Turkey Subfamily (Meleagridinae)
Wing: Short; rounded; concave; primaries curved.
Tail: Broad; rounded.

Verdin (Remizidae)
Wing: Rounded; ten primaries.
Tail: Rounded; shorter than wing.

Vulture, New World (Cathardidae)
Wing: Long; broad; rounded.
Tail: Short.

Waxwing (Bombycillidae)
Wing: Long; Pointed; ten primaries; outermost primary very short (only half the length of primary coverts).
Tail: Shorter than wing; square to rounded; upper tail coverts greatly extended.

Woodpecker (Picidae)
Wing: Moderately long; pointed; outermost primary very short.
Tail: May be pointed, rounded, or graduated; rectrices acuminate.

Wood Warbler Subfamily (Parulinae)
Wing: Varies in size and shape; usually long and pointed.
Tail: Usually shorter than wing; square to rounded.

Wren (Troglodytidae)
Wing: Short; rounded; concave; ten primaries.
Tail: Varies in size; rounded; rectrices soft with rounded tips.

ELEVEN

Position and Use

A bird's tail has many uses and can be seen in various positions. The most practical use of the tail is during flight. The surface area of the tail adds to the surface area of the wings and creates the additional lift necessary for flight. The tail also becomes a steering rudder, enabling a flying bird to turn left or right, or an elevator controlling the bird's up or down flight. A study of how birds fly is necessary to fully understand the functions of the tail during flight.

The tail is often used as an air brake to slow down forward speed when the bird is landing. As a bird is about to alight, the tail is spread wide and lowered to increase resistance to the air. This helps to slow down forward speed so that there is not a jolt when the bird alights on a perch. At the same time, the wings are used for the same purpose. The tail is also a stabilizer to help maintain a bird's balance while it is perched.

The tail of a bird is predominant in many of the behavior patterns described previously. Nearly everyone is familiar with the fanned-out tail of the male turkey during its mating display. Renderings of the ruffed grouse are often depicted in the same manner. Some species use the tail as a signal flag in what is called flight intention movements. Just before taking off, the bird flicks its tail as a signal, as is seen in the dark-eyed junco, which flashes its white outer tail feathers.

The tail of a few species is used as a tool, as is seen in the woodpecker, which uses it as a prop while it clings to the side of a tree. This constant use of the tail causes the tips of the rectrices to wear out and become frayed, sometimes to the point of making the tail shape nearly unrecognizable.

TWELVE

Coloration

The colors of a bird are what makes it appealing to most people. Although birds spend a lot of time preening, beauty is not the primary purpose for the vast array of colors found on the various species, except in some males during the mating season. The colors of a bird are primarily present as a means of camouflage. For the purpose of nesting and incubation, camouflage is especially important in the female of any species. The second most important purpose of coloration is, perhaps, for species identification. Coloration appears to be a primary method for birds of the same species to identify each other, thus preventing interspecies mating.

On the wings and tail of a bird, colors are produced in the feathers by two basic means—pigment coloration and structural coloration. Pigment coloration is color created by the presence of certain color pigments in the feather. There are three main color pigments found in the feathers of birds. Carotenoids are a type of pigment that produce intense reds, yellows, and oranges, as seen in cardinals and goldfinches. Porphyrins are a type of pigment that produce tones of browns, reds, and greens. Porphyrins are what give many owls their brownish colors. The most common type of pigment is melanin. Melanins produce the blacks, grays, and browns found in nearly all birds.

Structural coloration is the appearance of colors created by light being diffused, reflected, or diffracted as it strikes the structural surface of the feather. Structural coloration can be compared to a prism that can produce the colors of a rainbow when sunlight strikes it. This is the same process that produces structural colors of a bird. Iridescence is a form of structural coloration, as seen in the gorget of a hummingbird or the speculum on the wing of a duck. The speculum is a brightly colored, usually iridescent patch found on the secondaries of most ducks. Its unique name is derived from a Latin word meaning mirror.

Many times the colors on a bird are produced by a combination of pigments and structure. The green magpie, for example, achieves its green color as a result of a yellow pigment overlying a blue structural color (blue and yellow produce green). Since yellow pigment fades quite rapidly in sunlight, members of the green magpie that reside in sunny territory appear to be more bluish in color.

Often a single feather will have more than one pigment and structure on different areas of the feather. This combination of two or more pigments or structures on a single feather accounts for the different colored barring, spotting, vermiculation, tips, or vanes.

Pigments and structural coloration exist over the entire bird. Iridescence, though, is seldom seen in the tail except in a few species, such as the peacock. Tail color is often an identifying feature for bird-watchers. In some instances it is the feature for which a bird was named, as in the redtail hawk.

As in the wings, tails do not have just one hue or color tone. Many tails are marked with barring or bands. The rectrices of some species are tipped in white, while others have a white outermost

feather on each side of the tail. The coloration of immature birds is usually quite different than in an adult bird. The bald eagle, for example, does not have a white head and tail until its fifth year of growth. Color patterns on tail feathers often help identify the approximate age of a species and in many instances the sex of the bird.

Sometimes, two birds of the same species will appear in two different color phases. These color differences are not an abnormality. The eastern screech owl and ruffed grouse, for instance, appear as either gray or red, and these differences are referred to as color morph. When a species appears in only two color-morphs, it is a condition called dichromatism. When there are three or more color phases the condition is called polychromatism. The gyrfalcon is polychromatic and appears in three phases: white, gray, and black. Color changes often appear when the bird molts its plumage and are most obvious in the males during breeding season.

Occasionally, color differences in a species are abnormalities often caused by diet or disease. Abnormalities of pigments appear to be genetic, since they appear consistently with each generation. The presence of more or less than the normal amount of a pigment is called heterochromatism. Melanism is an excess of black or brown pigments, while flavism is an excess of yellow pigment. Too little of all pigments is known as dilution, or leucism, while the absence of all pigments produces albinism.

Perhaps one of the most outstanding conditions of abnormal plumage is gynandromorphism. In this condition of abnormality, all the feathers on one side of the bird are male colored while the other side is female in color. This condition has been found in pheasants, falcons, parrots, and woodpeckers.

Birds with abnormal plumage appear to be poorly suited to survival in the wild. They are more conspicuous to predators and are less likely to breed since they are not recognized by their own species.

Apparent color change, though, is not always from molting or abnormalities. Many times color change is the result of feather wear. House sparrows and snow buntings have light-colored feather tips during the winter but appear to be totally black during the breeding season. This occurs by the wearing away of the feather tips, leaving only the dark-colored area. After mating, these birds molt and grow fresh feathers that are tipped in a lighter color.

This same feather wear occurs in the herring gull, which has primaries tipped with black. The pigments in the black tip

Color change sometimes takes place because of feather wear, as seen in the snow bunting.

The unpigmented white tips of a herring gull's primaries wear away quickly, causing the feathers to then have black tips.

make that area more wear resistant than the white areas that have no pigments.

Every carver or artist knows that there are literally dozens of hues and tones found on any bird and good reference is a must when painting any species. Sources of good color reference, though, are not always readily available. Fresh taxidermy mounts provide a good color reference, as do study skins. Good-quality photographs can also be a valuable aid in selecting the paint colors. Published photos in books and magazines, however, are not always accurate because of the inks used and the printing processes.

Care must also be taken in selecting the proper colors and combinations for the composition. For example, depicting a male goldfinch in full breeding plumage with a winter scene might look pretty but would not be an accurate rendering of the species because breeding occurs in the spring. A zoo, aviary, rehab center, or your backyard is the best place to observe the correct colors of the species you are depicting.

THIRTEEN

Learning More

For the serious wildfowl carver and artist, there is a never-ending search for information. This search began, scientifically, with the Greek philosopher Aristotle, who recognized over one hundred different species of birds and recorded information about them. Today, more than two thousand years later, the search for information goes on.

For more than fifteen thousand years, birds have been depicted in art, beginning with the cave wall paintings during the Stone Age. Ancient Egyptians used bird shapes to depict their many gods. Toth, the god of wisdom, had the head of an ibis, and Horus, the sky god, had the head of a falcon. Myth, legend, and folklore have included birds since the beginning of mankind.

Many of the early scientific studies of birds were done by people who were primarily artists. All these artists attempted to bring accuracy and authenticity into their work. Today, perhaps one of the best known and most highly contributive wildfowl artists is Roger Tory Peterson. His field guides, with key identifiers, have become a part of every bird-watcher's collection of informative bird literature.

For the serious wildfowl artist, whatever the medium—carving, sculpting, casting, painting, illustrating—there is the need to constantly acquire more information about the subject. The wildfowl artist can learn more in many different ways. One way, of course, is by reading publications such as this, but many other channels of learning are available.

Bird-watching field trips often produce abundant sources of information, and joining the activity of bird banding can provide close-up examination of certain species. For the artist who is not capable or willing to be active in the outdoors, there are many other available sources of information in the form of videotapes, study skins, photographs, taxidermy mounts, rehabilitation centers, and zoos.

Self Test

1. The wing of a bird is made up of _____ segments.

2. The outermost segment of the wing is called the _____.

3. All birds have ten primary feathers. (True or False)

4. The primary feathers of a bird are all shaped the same. (True or False)

5. All the primaries and secondaries together are called _____.

6. The scapulars grow in what area of the wing?

7. The two or more prominent feathers growing from the alular digits are called _____.

8. The soft coverts on the underside of the wing are referred to as the _____.

9. Mantling is a display behavior used by _____.

10. The only use of a bird's wing is for flight. (True or False)

11. The periodic shedding and replacement of feathers is called _____.

12. The tail of a bird has only one bone, called the pygostyle. (True or False)

13. The crissum is the correct terminology for the _____.

14. The uropygial gland is more commonly known as the _____.

15. The colors of any one species are always the same. (True or False)

16. The speculum is a bone of the wing skeletal system. (True or False)

17. What are the rectrices?

18. The largest, most powerful muscle of any bird that is capable of flight is the _____.

19. The feathers growing from the armpit area of the wing are called _____.

20. All the feathers we see on the surface of a bird are called _____.

Answers on page 88

Glossary

Abductor. Any muscle that causes a limb segment to move away from the longitudinal centerline of the body; see also EXTENSOR.

Accipiter. Name given to a group of short-winged hawks, which include the goshawk, sharp-shinned hawk, and Cooper's hawk.

Action pattern. The lowest level of behavior pattern that is a motor action, such as flapping flight.

Acuminate. A feather shape where the tip comes to a sharp point, as in woodpeckers.

Acute. A tail shape where the center rectrices are the longest, as in the ring-necked pheasant; pointed.

Alar tract. The main feather tract of the wing, both dorsally and ventrally.

Albinism. The absence of color pigments in a species that results in a white color when the normal is not white.

Alula. Collectively, the group of feathers made up of the alular quills that emanate from the first digit of the wing.

Alular digit. The first digit of the manus; corresponds to the thumb of a human.

Alular quill. Any one of the feathers emanating from the alular digit.

Antebrachium. The middle segment of the wing; forearm.

Anterior. Referring to the front or frontal portion.

Apteria. Featherless areas of the skin between feather tracts.

Articulated. A type of joint where the adjoining bones can move with respect to each other.

Axilla. The somewhat hollow space on the ventral side where the wing joins the body; the armpit.

Axillaries. See AXILLARS.

Axillars. A group of prominent feathers growing from the armpit area of the wing.

Bastard wing. Colloquial term for the alula.

Behavior pattern. A repetitive motion or action by a species that constitutes a particular behavior.

Bend of the wing. Common terminology for the wrist joint.

Brachium. The segment of the wing closest to the body containing the humerus.

Breeding plumage. See NUPTIAL PLUMAGE.

Brevis. From a Latin word meaning brief or short.

Broad. A wing shape where the primaries, and particularly the secondaries, are relatively long, as in the sharp-shinned hawk.

Buteo. Name given to a group of hawks that share similar characteristics. They have long, broad, rounded wings. Includes red-tailed, red-shouldered, and rough-legged hawks. Collectively they are called buteos.

Cardiac. Pertaining to the heart; a type of muscle found only at the heart.

Carotenoid. A type of color pigment in birds that produces orange and yellow colors; named after the common carrot.

Carpal remix. A shorter, secondary remix located in the gap between the primaries and secondaries.

Carpals. Small bones of the wrist joint.

Carpometacarpus. The large, fused bone of the manus.

Caudal tract. The feather tract of the tail that is located both dorsally and ventrally.

Caudal vertebrae. The last six vertebrae of the spinal column, which forms the base of the tail; see also PYGOSTYLE.

Clavicle. One of three bones that make up the pectoral girdle.

Concave. A wing that has an obvious extreme curve of the surface plane, as in the ruffed grouse.

Contour feathers. All the feathers visible on the surface of a bird, including the flight feathers. Contour feathers give a bird its contour or shape.

Coracobrachialis. One of the muscles of the breast that assists in drawing the humerus toward the body.

Coracoid. A short, thick bone of the pectoral girdle.

Crissum. Collectively, all the underside tail coverts.

Cubitals. Technical terminology for the secondaries, derived from the Latin word *cubitum*, meaning elbow.

Dermis. The underlayer of the two layers of the skin.

Diastataxic. A condition where the fifth secondary feather is absent; see also EUTAXIC and DIASTEMA.

Diastema. The gap between the fourth and sixth secondary feathers of a diastataxic bird; see also EUTAXIC.

Dichromatism. A condition where a species appears in two different color phases. The eastern screech owl appears in a red phase and a gray phase. See also POLYCHROMATISM.

Digital primaries. The primary feathers that are attached to the digits of the manus; see also METACARPAL PRIMARIES.

Digits. Referring to the toes or fingers. The wing of a bird has three digits. See also ALULAR DIGIT, MAJOR DIGIT, and MINOR DIGIT.

Dilution. See LEUCISM.

Distal. In ornithology, the point or segment farthest from the longitudinal centerline of the body.

Dorsal. Referring to the top side or upper surface.

Drumming. A mating behavior of the male ruffed grouse where the wings are rapidly beat to produce a staccato drumming sound.

Elbow. The joint of the wing formed where the humerus meets the ulna.

Emarginate. A tail shape where the rectrices increase in length from the center to the outside in successive gradations.

Epidermis. The outermost surface layer of the two-layered skin.

Ethology. The scientific study of animal or bird behavior.

Eutaxic. When there is one greater secondary covert for each secondary feather, the bird is said to be eutaxic, as opposed to diastataxic.

Extensor. Any muscle that extends or straightens a joint.

Facia. Thin tissue surrounding bundles of muscle fiber.

Falco. Name given to the group of falcons collectively. Birds in this group have long, narrow, pointed wings that are swept back.

Feather follicle. The socket in the skin from which a feather grows.

Feather tract. An area on a bird where a particular group of feathers grow; e.g., caudal tract, alar tract, etc.

Flat. A wing shape that has little or no surface plane curvature, as in a hummingbird.

Flavism. A condition where there is an excess of yellow pigments in the bird.

Flexor. Any muscle the causes a joint to bend to a lesser angle.

Flight feathers. Collectively, all the large, stiff feathers of the wing and tail that pertain primarily to flight; the primaries, secondaries, and rectrices.

Forearm. Common terminology for the middle segment of the wing containing the ulna and radius; see also ANTEBRACHIUM.

Foramen triosseum. The opening in the pectoral girdle formed where the clavicle joins the coracoid and scapula.

Forcula. The fused lower ends of the two clavicle bones; the wishbone.

Forked. A tail shape where the outermost rectrices are the longest and the remaining rectrices gradate successively to the center.

Glenoid cavity. The socket of the ball-and-socket joint for the head of the humerus on the pectoral girdle.

Gorget. The iridescent bib of a hummingbird.

Graduated. A tail shape where the centermost rectrice is the longest and the remaining rectrices gradate successively, in steps, to the outermost.

Greater primary coverts. Topographically, the first row of contour feathers that cover the bases of the primaries. There is one greater primary covert for each primary feather.

Greater secondary coverts. Topographically, the first row of contour feathers that cover the bases of the secondaries; see also EUTAXIC and DIASTATAXIC.

Gynandromorphism. A rare condition where the bird has male coloring on one side and female coloring on the other side.

Harrier. Name given to a group of hawks that share similar characteristics. Wings are long and nearly pointed with a long, slender tail. Harriers, such as marsh hawks, usually fly low to the ground in search of prey.

Head. The enlarged proximal end of the humerus that is the ball of a ball-and-socket joint on the pectoral girdle.

Heterochroism. The presence of too much, or too little, of a color pigment.

Humeral tract. Feather tract along the dorsal side of the humerus. Considered by some authorities to be a separate tract, it is part of the scapulohumeral tract.

Humeral patagium. The fold of skin that extends from the front of the elbow to the shoulder along the humerus, to form the armpit.

Humerus. The rather short, thick bone of the innermost segment of the wing.

Integument. The outer covering of a bird or animal, which includes the skin, hair, feathers, and scales.

Joint. The point where two or more bones meet.

Keratin. A protein substance that forms the feathers and claws on a bird, and the nails in humans.

Lesser primary coverts. Topographically, the forward-most row of coverts on a wing that overlap and cover the median primary coverts. Not all species have a row of median primary coverts, in which case the lesser primary coverts overlap the greater primary coverts.

Lesser secondary coverts. Topographically, the forwardmost rows of coverts on a wing that overlap and cover the bases of the median secondary coverts; often cannot be distinguished from the marginal coverts.

Leucism. A condition where there is too little of all the color pigments; See also DILUTION.

Ligament. Strong fibrous tissue that binds the bones together at a joint.

Little remix. See REMICLE.

Long. A wing size where the wing is equal to or greater than the length of the trunk excluding the head and tail.

Longus. Latin term meaning long; used in naming muscles.

Major. In muscle nomenclature, refers to the larger of two similar muscles.

Major digit. The largest, outermost digit of the wing.

Mantling. A display behavior used by most birds of prey where the wings are partially spread and drooped over the freshly killed prey.

Manus. The outermost segment of the wing, corresponding to the human hand.

Marginal coverts. The small contour feathers that cover the forward margins of a wing; see also WING LINING.

Mating display. A ritual behavior by the male of a species to attract a mate. Mating displays are markedly different from precopulatory displays.

Median primary coverts. Topographically, the second row of feathers that cover the bases of the primaries and overlap the greater primary coverts.

Median secondary covert. Topographically, the second and sometimes third and fourth rows of feathers that cover the bases of the secondaries.

Medullary bone. A spongy material produced in the pneumatic cavities of the bones of female birds. This occurs during the reproductive cycle and is a means of storing extra calcium needed for producing eggshells.

Melanin. A type of color pigment that produces blacks, grays, and browns.

Melanism. A condition where there is an excess of melanin.

Metacarpal primaries. The primary feathers that are attached to the metacarpal bones; see also DIGITAL PRIMARIES.

Metacarpals. Bones of the manus that are fused together to form the carpometacarpus.

Minor. In muscle nomenclature, refers to the smaller of two similar muscles.

Minor digit. The smallest of the three digits of the manus.

Molt. The loss and replacement of the feathers. Most birds go through two molts a year.

Morph. When a species appears in two or more different colors, as in the ruffed grouse, which appears in a red or gray morph.

Myology. The scientific study of muscles and muscle systems.

Narrow. A wing shape where the primaries and secondaries are short relative to the length of the wing.

Nuptial plumage. The plumage that is worn, usually in the spring, prior to mating; the breeding plumage.

Oil gland. See UROPYGIAL GLAND.

Osteologist. One who scientifically studies bones.

Passerine. Referring to the largest order of birds, the Passeriformes, which include mainly birds of perching habit.

Patagium. The fold of skin on the wing between the wrist and the humerus.

Pectoral girdle. In the skeletal system, the bones that form the shoulder of the bird and support the wing.

Pectoralis. The largest, most powerful muscle of the wing, which forms the breast of the bird.

Phalanges. The bones of the digits, either in the toes or the manus.

Phalanx. Singular of phalange.

Pigment coloration. Color produced solely by a pigment.

Pinion. Term sometimes used to identify the hand segment or primaries collectively.

Pinioning. The act of removing the hand segment and primaries to prevent flight.

Pisolunare. One of the carpal bones in the wing of a penguin; see also SCAPHOLUNARE.

Pointed. See ACUTE.

Pollex. Early terminology for the alular digit.

Polychromatism. When a species appears in more than two color phases; e.g., the gyrfalcon appears as white, gray, or black.

Porphyrins. Color pigments that produce tones of browns, reds, and greens.

Postcopulatory display. In some species, a ritual display performed immediately after copulation by either one, or both, of the sexes.

Posterior. Referring to the rear or hind part or segment.

Precopulatory display. In many species, a ritual display performed prior to copulation; see also MATING DISPLAY.

Preen gland. See UROPYGIAL GLAND.

Primaries. Large, stiff, flight feathers that emanate rearward from the posterior edge of the manus.

Pronator. Any muscle that causes a segment to rotate forward around the axis of the joint.

Proximal. Referring to the point closest to the longitudinal centerline of the body.

Pterylae. The areas of a bird from which feathers grow, as opposed to apteria.

Pygostyle. The fused, last vertebrae of the spinal column that is the support for the rectrices.

Quill knobs. Small protrusions, or bumps, often found along the posterior edge of the ulna where the secondaries are attached.

Radiale. A small, squarish-shaped bone of the wrist.

Radius. The more slender of the two bones of the antebrachium, or forearm.

Rectrices. Collectively, the large flight feathers of the tail; see also RECTRIX.

Rectrix. Singular of rectrices; one flight feather of the tail.

Remicle. The reduced-size, outermost, primary feather found in some species; also called the little remix.

Remiges. Collectively, all the flight feathers of the wing, including the primaries and secondaries.

Rounded. A wing shape where the middle primaries are the longest and the remaining primaries gradate successively to either side, giving the open wing a round shape, as in the quail or grouse; also a tail shape where the middle pair of rectrices are the longest and the other rectrices are successively shorter on each side, as in the tail of a crow.

Scapholunare. A small wrist bone of a penguin.

Scapula. One of three bones that make up the pectoral girdle.

Scapulars. A group of prominent feathers growing in the scapular region of the wing.

Scapular tract. A feather tract on the shoulder of a bird. Considered by some authorities to be a separate tract, it is part of the scapulohumeral tract; see also HUMERAL TRACT.

Scapulohumeral tract. Feather tract of the wing located in the humeral and scapular areas.

Secondaries. The large, stiff, flight feathers emanating rearward from the posterior edge along the middle segment of the wing.

Sesamoid bones. Two small bones found in the elbow of a penguin.

Shaft. The quill that extends the length of a feather near its centerline.

Short. A wing size that measures less than the length of the trunk, excluding the head and tail.

Smooth. A type of muscle found mainly in the skin of a bird.

Speculum. An iridescent patch of color on the secondary feathers of most ducks.

Spinal column. Collectively, all the bones that make up the backbone of a bird or animal.

Spinose. A feather shape where the quill, or shaft, extends at the tip without barbs along the sides.

Square. A tail shape where the rectrices all end in the same length.

Sternum. The breastbone of a bird.

Striated. A type of muscle made up of bundles of muscle fiber; see also CARDIAC and SMOOTH.

Structural coloration. Color produced solely by a certain structure of a feather. Iridescence and white are forms of structural coloration.

Supracoracoideus. A muscle of the breast located beneath the pectoralis muscle; see also CORACOBRACHIALIS.

Suture. A type of joint where the adjoining bones do not move with respect to each other.

Synovial. A type of joint totally encapsulated by a tissue that is filled with a fluid for lubrication.

Synovial fluid. The lubricating fluid within a synovial joint.

Tendon. Strong fibrous tissue that connects a muscle to the bone.

Tertials. The feathers that grow from the humeral segment of the wing on the dorsal and ventral side; normally longer and stiffer than the coverts; the humeral feathers.

Tertial tract. Feather tract or region of the wing along the humerus.

Tertiaries. See TERTIALS.

Topography. In ornithology, the mapping and labeling of the surface area of a bird.

Ulna. The thicker of the two bones of the antebrachium, or forearm.

Ulnare. A small, squarish-shaped bone in the wrist or bend of the wing.

Under tail coverts. See CRISSUM.

Upper tail coverts. The group of contour feathers that cover the bases of the rectrices.

Vent. An opening on the underside of a bird where digestive and other wastes are excreted.

Ventral. Referring to the bottom or underside of an area.

Vertebrae. Collectively, the bones that make up the vertebral column; see also SPINAL COLUMN.

White fibers. One of two types of tissue found in muscles; the other type is red fibers. Birds that are not strong fliers tend to have more white than red fibers in the breast muscles.

Wing claws. A claw that emanates from the first, and sometimes the second, digit of the manus, as in the hoatzin.

Wing lining. Term used to describe all the coverts, collectively, on the underside of the wing.

Wrist. The most distal joint of the wing, formed where the manus meets with the ulna and radius; the bend of the wing.

Bibliography

Although a great number of sources were used as reference material for this book, the primary sources were as follows:

Brooke, M., and T. Birkhead, eds. *Cambridge Encyclopedia of Ornithology.* New York: Cambridge Press, 1991.

Farner, D. S., J. R. King, Jr., and K. C. Parkes, eds. *Avian Biology,* Vol. II. New York: Academic Press, 1971.

Pettingill, O. S., Jr. *Ornithology in Laboratory and Field.* Orlando, Fla: Academic Press, 1985.

Proctor, Noble S., and Patrick J. Lynch. *Manual of Ornithology.* New Haven: Yale University Press, 1993.

Sturkie, Paul D. *Avian Physiology.* Ithaca, N.Y.: Cornell University Press, 1965.

VanTyne, J., and A. J. Berger. *Fundamentals of Ornithology.* New York: Dover Publications, 1971.

Wilson, Barry W. *Birds, Readings from Scientific American.* New York: W. H. Freeman Co., 1980.

TEST ANSWERS

1. Three

2. Manus or hand

3. False. Some species have only nine, while others have eleven or twelve with ten being most common.

4. False. The primaries of all birds vary slightly in shape.

5. Remiges or flight feathers

6. Shoulder area

7. Alula, alular quills, or bastard wing

8. Wing lining

9. Birds of prey (hawks and owls)

10. False. Wings are used in displays, distractions, protection, and swimming, as well as for flying.

11. Molting

12. False. The tail of a bird contains six caudal vertebrae as well as the pygostyle.

13. Under tail coverts

14. Oil gland or preen gland

15. False. Often the male is a different color than the female of a species, and juvenile birds are usually differently colored than the adults.

16. False. The speculum is a patch of bright color on the wing of a duck.

17. The flight feathers of the tail

18. Breast muscle, or pectoralis

19. Axillars

20. Contour feathers